PELI

THE FACE

G. H. Dury, after accelerated undergraduate progress, obtained his London B.A. in geography in 1937. He spent most of the Second World War, including a year on loan to the U.S.A., in R.A.F. Photo-Intelligence. His main subsequent employment has been at Birkbeck College (London); at Sydney University, Australia, where he was McCaughey Professor and Head of Geography, and also Pro-Dean and then Dean of Science; and at the University of Wisconsin-Madison, where he is Professor of Geography and Geology/Geophysics, and has served as Geography Chairman. He has had charge of Geography and Geology at a technical college, has been attached to the U.S. Geological Survey as Division Staff Scientist (Water Resources), has taught at Cambridge, and has been Visiting Professor of Geology at Florida State University, Tallahassee. His fieldwork in the British Isles, Sweden, Australia, the U.S.A., and Brazil is supplemented by extensive additional travel. His publications include numerous technical papers, several monographs, books on *Map Interpretation*, *The British Isles*, *The United States and Canada* (with R. S. Mathieson), edited collections on geomorphology, rivers and river terraces, the geography of Australia (with M. I. Logan), plus contributions to the *Encyclopaedia Britannica* and its Year-book series. He serves on the editorial boards of four technical journals. His M.A. and Ph.D. degrees from London were followed in 1971 by the D.Sc., awarded for published research. In 1975 he received the Meritorious Contribution Award of the Association of American Geographers. His current interests include reading in four languages, glacial effects on landscape, stream channel geometry, the geomorphic effects of climatic change, regional description, good technical writing, and the production of successors who will outrun him.

*G. H. Dury*

# THE FACE OF THE EARTH

FOURTH EDITION

PENGUIN BOOKS

Penguin Books Ltd, Harmondsworth, Middlesex, England
Penguin Books, 625 Madison Avenue, New York, New York 10022, U.S.A.
Penguin Books Australia Ltd, Ringwood, Victoria, Australia
Penguin Books Canada Ltd, 41 Steelcase Road West, Markham, Ontario, Canada
Penguin Books (N.Z.) Ltd, 182–190 Wairau Road, Auckland 10, New Zealand

—

First published 1959
Reprinted 1960, 1962, 1963, 1964
Second edition 1966
Reprinted 1968, 1969, 1970
Third edition 1971
Reprinted 1973, 1974
Fourth edition 1976

—



—

Made and printed in Great Britain
by Cox & Wyman Ltd, London, Reading and Fakenham
Set in Monotype Times

# CONTENTS

# LIST OF PLATES

# LIST OF TEXT FIGURES

# ACKNOWLEDGEMENTS

I AM grateful to D. G. A. Whitten and G. T. Raine for having suggested that I write this book, and for continual encouragement during composition. Mr Raine, R. H. C. Carr-Gregg, and M. H. Dury have kindly commented on the text in draft. Luna B. Leopold, of the U.S. Geological Survey, has helped in a less immediate but no less important way, by supplying most illuminating information on the behaviour of rivers. Responsibility for errors and inaccuracies remains my own.

Part of the text relating to glacial breaches, in Chapter 14, has been adapted by permission from an article of mine in *Science News* 38, from which Figs. 79, 80, and 81 are also reproduced.

Those photographs which have not been taken by myself have been supplied by the following, to whom thanks are due for permission to use them here:

*The Times* for Plates 2 and 67; the High Commissioner for New Zealand for Plate 4; the United States Information Services for Plates 5, 13, 21, 24, 44, and 80; the Controller, H.M.S.O., for Plates 7, 22, 27, 31, and 49; Aerofilms Ltd for Plates 12, 23, 25, 26, 33, 36, 47, 48, 78, 79, and 81; R. Kay Gresswell for Plates 19 and 64; the Australian News and Information Bureau for Plate 28; Photoflight Ltd for Plates 29 and 45; the Yugoslav Tourist Office for Plate 30; V. C. Browne for Plate 38; the Irish Tourist Agency for Plates 41 and 70; Canadian National Railways for Plates 42, 54, 56, and 62; the Swiss National Tourist Office for Plates 53 and 58; Canadian Pacific Railway for Plate 55; the Norway Travel Association for Plate 63; the Finnish Embassy for Plates 68, 72, and 73; Wm S. Thomson for Plate 74; and Spence Air Photos for Plate 82. Plates 51, 52, and 66 were specially taken for this book by Sylvia Hall.

G. H. D.

*London*
*May* 1959

# CHAPTER ONE

## *Introductory*

THIS book is intended to present land-forms to non-specialists. It is an outline of the sculpture of the earth's surface by natural processes.

Land-forms evolve. Each piece of country has its own history of erosion. Measured on the geological time-scale, some such histories are very brief, but others attain the respectable lengths of many millions of years. Some are fairly simple, others almost incredibly complex. Each, however, is systematic and sequential – in any one locality, physical landscapes follow one another in orderly succession.

Almost any piece of country includes remnants of former landscapes, as if it were a document with scraps of past history written on it. Some of the earlier notes are obscured by the later, and parts of the record are inevitably missing. All the skill of a logician – or of an amateur detective – is required in the work of decipherment. At times, the task of reconstruction is as formidable as the restoring of a whole film from a single photographic still.

The formal name of studies in earth-sculpture is geomorphology – knowledge of the shape of the ground. Although geomorphology is one of the youngest natural sciences, it already has a technical vocabulary of its own. But while technical terms are useful, proper, and necessary in technical writing, they are used as sparingly as possible in the following text, both for the sake of simplicity in and order to save the space required by definitions; they appear only where they prevent cumbrous paraphrase. Diagrams and photographs, on the other hand are freely employed, even at the expense of text, for the material to be discussed is essentially visual.

For all its growing vocabulary and yeast of ideas, the systematic study of land-forms has barely reached full scientific status. In some respects, it stands today where botany and zoology stood in the eighteenth century and where geology and archaelogy stood in the nineteenth. Anyone committed to geomorphology is to a large extent an observer, and a pioneer observer at that. No more than a small fraction of the world's land-surface has yet been closely examined. Field-workers can rely on making new, exciting and valuable discoveries. No doubt this circumstance partly accounts for the prevailing enthusiasm.

Just because there is so much to do, progress can be swift. It is greatly helped by workers on related topics. Important contributions come from geologists – indeed, geomorphology overlaps widely with geology, from which it is in part derived. Some of the greatest geomorphologists are geologists by profession. Others are geographers, interested mainly in the physical side of their subject. But the emphasis of geomorphology proper differs from the emphasis of geology or of geography. Most geologists are concerned with the formation and deformation of rocks, geomorphologists with the erosional facets by which rocks are truncated. Geographers as a group deal with man in relation to his environment. As specialists, geomorphologists examine certain aspects of that environment, without necessarily considering man at all.

Enough is already known about the form of the ground for geomorphology to claim separate identity. Naturally enough, demands for independence provoke suspicion and distrust. A few geologists regard geomorphology as superficial – in more than one sense. Some geographers, alarmed by the evident vigour of geomorphology, react against the seduction of wholly physical knowledge. But although, in the process of gaining separate status, geomorphology seems to be detaching itself from geology on the one side and from geography on the other, its interests tend in practice to merge with the interests of work passing under other names. No one with an eye for country can fail to perceive some kind of relationship between the form of the ground, the nature of underlying rocks, the character of soil, the types of wild plants, and the location and quality of cultivated land. Complex interrelationships of the kind implied here are recognized in practical fashion wherever land quality and land potential are being deliberately assessed.

Undeveloped lands are being studied in Australia and the U.S.S.R.; applied regional planning in the U.S.S.R. is matched in the U.S.A. to a greater or lesser extent by floodplain zoning, regional design, landscape architecture, and landscape engineering. Problems and possibilities in the U.K. differ somewhat from those of countries where vast extents of open space still exist. On the other hand, the current and mounting concern with environmental damage, control, management, and conservation involves geomorphology equally with a whole additional range of natural and applied sciences. To name a few, these include geology, hydrology, soil mechanics, civil engineering, pollen analysis, and radiometric dating. Whether by overlap or by enforced borrowing, geomorphology is acquiring a range of experimental techniques, and is developing a general theory of its own. Simultaneously, the constant re-interpretation of actual landscapes continues to produce new and exciting concepts. Accordingly, the following chapters include accounts of recent work alongside statements of well-established principle, and reference to quantitative studies in addition to generalizations.

No book of this size can do justice to my subject, to my predecessors, to my teachers, or to my fellow-workers. Some of the omissions enforced by lack of space are, I hope, offset by the lists of references. The general risk of injustice is, perhaps all the greater because I have neither avoided controversial topics nor hesitated to commit myself to one side of a dispute where the evidence seems to warrant a choice. As A. J. Ayer says in another context, 'It is doubtful if there are any facts about which no one could be mistaken'. I shall suggest that very grave mistakes can be, and have been, committed in the interpretation as well as in the observation of the facts of landscape. But if geomorphology were so easy that nobody could be wrong, it would scarcely be worth studying.

Many of the examples chosen to illustrate the character of surface features and the work of erosional processes are those which I know at first hand, or which strike me as particularly instructive. In this respect the book is unashamedly partial. It seems however to prove reasonably attractive, appearing in the list of the hundred best Pelicans, and enabling me to share with Mr Salteena in *The Young Visiters* a taste for fresh air and royalties. That is to say, it helps to sustain my fieldwork. Its main purpose continues to be that of encouraging readers to look at scenery with new eyes.

CHAPTER TWO

# *Rock-Destruction*

THROUGHOUT this book, emphasis will be laid on the destruction and removal of the rocks of the earth's crust. We are concerned primarily with earth-sculpture by the agencies of weathering and erosion. But processes of deposition deserve at least a passing mention, not only because they are complementary to erosion, but also because it is desirable to forestall any dismal impression that the world's lands are doomed eventually to disappear entirely.

Material stripped off the lands eventually finds its way to the sea. Its movement is irregular, and it accumulates in distinctive constructional forms from place to place – for example, as deltas on the shores of inland lakes, as steep banks of scree on the flanks of mountains, or as spreads of alluvium in valley-bottoms. In the end, however, it is carried into the oceans, and goes to form new rocks on the sea-bed.

A portion of the total bulk – the coarsest material – settles on the sea-bed under the direct influence of gravity. A much larger portion, consisting of very fine grains, is brought down in suspension and can remain suspended for a time in sea-water; its microscopic particles are so tiny that they can be held up by water, just as very fine grains of dust can remain suspended in the air. But under the influence of sea-water the particles flocculate, forming sizeable crumbs which fall to the bottom. A third portion is brought down in solution. It tends to increase the salt content of sea-water, but not all of it remains in the dissolved state. Part is extracted by living things, being fixed in solid form as the bones of fish, the shells of molluscs, and the cups of corals, while part is directly precipitated by chemical reactions.

Marine sediments formed in one or other of these ways are transformed into firm rock. Water is driven out by the pressure of overlying sediments, and natural cements bind certain types of rock very firmly. If the sedimentary rocks are raised above the level of the sea, the processes of weathering and erosion begin to act on them, and the work of rock-destruction and rock-removal continues. If, on the other hand, the sediments are carried down to very great depths by subsidence of the crust, they may be melted. In that case they supplement the supply of molten rock which will eventually flow over the land-surface as lava, or will solidify deep underground. Even if it solidifies at depth, the rock can be exposed at the surface when erosion removes its cover. It can thus be seen that rock-destruction is merely one part of a continuous cycle, in which the materials of the crust can be used over and over again.

Land above sea-level can exist only if the forces of construction more than counterbalance those of destruction. It is 3000 or 4000 million years since a solid crust first began to form on the earth, and there is no reason to think that the land-masses of today are any less extensive than they were originally. In the long run, that is to say, the forces of construction have prevailed.

Erosion strips rock from the upstanding lands. Subsidence reduces their extent. The geological record shows that repeated subsidence has affected very large areas. But subsidence has alternated with uplift, which has counterbalanced subsidence in the long run. Although the height, shape, and extent of the continents have been changed many times, continents have existed ever since there has been a solid crust. Weathering and erosion have been at work throughout the same period.

In any event, the forces of destruction are not entirely adverse in their effects. If they did not operate, there would be very little soil. Consequently, there would be very little land-vegetation of familiar kinds, and animal life as we know it would be impossible. There would, of course, be no human beings.

This topic is worth a short discussion with the object of making clear the difference between the term *soil* and the term *waste-mantle*. Soil, in the biological sense, is the product of three sets of processes. First come the changes produced by the action of living things – plants, earthworms, soil bacteria, and the like. Then there are the changes due to the movement of soil-moisture. Fine particles are washed downwards into the lower levels of the soil. In dry conditions, when evaporation from the surface is taking place, dissolved substances can be carried upwards, but in wet conditions the movement of solutions is downward. Vertical movements of soil-moisture reinforce the changes produced by organisms, the general result being to arrange the soil in layers parallel to the surface of the ground (Plate 1). The third set of soil-forming processes provides the raw material from which soil is formed. This set includes all the processes of weathering, by which rock is broken into fragments and is chemically changed. Weathered material – broken and rotted rock – is known collectively as the waste-mantle. Soil may be roughly defined as that part of the waste-mantle which has been transformed by the combined action of soil-moisture and of living things.

Weathering, and the waste-mantle which it produces, constitute the main theme of the present chapter. Some of the results of weathering are familiar to all who live in towns where the atmosphere is polluted. Other results are clearly visible in quarry-sections and on the faces of cliffs. Still others are widespread in mountainous districts. In all such places, rock can be seen in process of rotting or of breaking.

In the great towns of western Europe, the signs of weathering are only too clear. Many ancient buildings, and some not so ancient, are rotting away (Plate 2). Their stones are fast crumbling in the dirty town air, beneath a layer of town grime. Blame for the rapid decay of architectural stone can rightly be laid on the smoky atmosphere, but the rocks of the earth's crust are also subject to rotting in natural conditions. They too are liable to become soft, to crumble, or even to be entirely dissolved away. Where the climate is too cold for rocks to rot freely, rock-breaking occurs. Each kind of climate promotes its own kind of rock-destruction.

The most spectacular results of weathering occur, in fact, where rocks are broken rather than rotted. Whereas the rotting of rocks produces a thick waste-mantle, which conceals from view the forms of unweathered rocks, rock-breaking produces jagged fragments which, falling away, reveal the fresh rock beneath (Plate 3) and accumulate in distinctive forms lower down the slope.

Many mountainsides in regions of cold climate are mantled with scree – great sheets of shattered, angular rock, resting at a high angle. Scree is typical of steep slopes where bare rock is exposed to the action of frost. It forms very freely where the rock is divided by numerous joints, so that pieces can be easily detached. When the surface temperature of the rock is above freezing-point, water percolates into fissures. When the temperature falls, and the water in the fissures freezes, great pressure is exerted on the enclosing rock, and thick flakes or sharp-edged blocks are wedged off. In some places the resulting scree takes the form of flat, sloping sheets, but if the falling fragments are guided by channels, the scree spreads out fanwise in half-cones. The scree shown in Plate 3 is part of such a half-cone, formed by the discharge of fragments down a narrow gully.

Beyond the bank of scree is seen a sheer rock-face, weathered out of alternating beds of limestone and shale. The shales are weaker than the limestones, so that the latter stand out in relief. Fantastic shapes can be produced by the selective attack of weathering – an attack frequently reinforced by surface wash (Plates 4, 5).

In the dry climates of hot deserts, rock-breaking is a more complex matter than in the frosty climates of high mountains. Generally speaking, weathering in deserts tends to break rocks down into their constituent grains. The end-product is frequently not scree, but sand. Since the sand can be blown away, rocky hills in the desert often stand up very abruptly above their surroundings (cf. Plates 78, 79).

Hot deserts are not only very dry – they are also sunny. By day, the sun blazes down, scorching the surface of the ground. By night, heat escapes freely into the cloudless sky. The temperature of the air near the ground varies greatly between day and night, and the surface-temperature of bare rock varies even more. As desert vegetation is sparse, and desert sand is mobile, much rock lies bare to the full force of the sun.

Most rocks are excellent insulators of heat. The daily changes of temperature do not therefore penetrate more than a few inches below the surface. Many rocks exposed in deserts scale away in thin flakes, or split away in thicker slabs. Flakes, slabs, and broken blocks disintegrate rapidly into smaller fragments or into their constituent grains. It is tempting to hold thermal effects responsible for the general breakdown. Expansion and contraction might be expected to tear grains of sandstone loose from their cement, or to disrupt crystalline rocks because of the different rates of expansion and contraction for different crystals. Dark crystals warm up and cool down more rapidly than do light-coloured crystals and crystals of contrasting size or composition exert contrasted pressures and tensions. But experimental work, whether in the laboratory or in the field, provides little support for the idea of thermally-controlled breakdown.

It is only fair to point out that experiments are so far limited, both in scope and in number. On the one side stand the facts that disintegration has produced enough sand to cover one-quarter of the world's area of hot desert, and that a considerable part of the remaining threequarters consists of bare (and disintegrating) bedrock or loose fragments. On the other side, the great bulk of desert sand consists of the chemically stable mineral quartz.

Solution in the desert environment, involving chemical action, is attested by the bitterness of lakes in desert centres, and by the shimmering expanses of saltpans. Few deserts in actuality are wholly rainless, and high daytime temperatures can be expected to promote vigorous chemical action when rain does fall. But, once again, some deserts are dry enough, at least in the short term, for dunes to be built not of quartz sand, but of the highly soluble precipitate gypsum, a common sediment of desert lakes. In this connection, as in so many others, we know far too little about what actually happens – except that disintegrated rock typifies hot deserts.

Specific effects of rock-rotting vary with the character of the rocks attacked. Complex chemical reactions occur, which are controlled partly by the climate and partly by the chemical make-up of the rocks. A thorough review of rock-rotting would require a discussion of rock chemistry – a subject which is outside the scope of this book. But a few examples can be given, which readily illustrate the kind of effect produced by chemical decay.

Solution is the simplest case. Rock-salt exposed at the surface, or in contact with water underground, is readily dissolved. Certain allied kinds of rock will also pass rapidly into solution at normal temperatures. Like rock-salt, they were originally precipitated from solution in the evaporating water of desert lakes. The high salt-content of spa-water, and the salinity of brine springs, are due to the dissolving of highly soluble rocks. Some of the water-holes of

Iraq are contaminated with solutions of Epsom salts, beds of which are being dissolved.

A slightly more complex instance is that of limestone, which is capable of rotting extensively (Plate 6). Limestone consists essentially of calcium carbonate ($CaCO_3$), which is insoluble in pure water, but can be dissolved by rain-water containing a minute proportion of atmospheric carbon dioxide in solution. On the human time-scale the decay of limestone is very slow, but on the long scale of geological time it is impressively rapid. Its general result is the complete dissolution of the rock. In intermediate stages, elaborate land-forms are produced.

Iron compounds occur in very many rocks. Like manufactured iron, they are attacked by rust. The brown colour of much beach-sand and of many sandstones is due to the rusting of natural iron. Other forms of chemical weathering usually accompany rusting, but at least one example due mainly to rusting can be quoted. It concerns the rock dolerite, which occurs in fissures in the earth's crust. The rock was injected into the fissures in a molten state, solidifying in flat sheets varying in thickness from a few inches to many feet. In the unweathered state, dolerite is a tough, compact rock, very dark in colour, composed of medium-sized crystals which include abundant crystals of iron compounds. Sheets of dolerite are typically traversed by systems of joints, which separate the rock into blocks and allow water to penetrate it.

When water attacks the iron compounds, the faces of the blocks rot and become soft. Dark, coherent rock is changed into rusty-brown, crumbly material. As the sharp edges and corners of the blocks are softened, the angular blocks are reduced to round boulders, each enclosed by its rotten shell (Plate 7).

It is true of rocks in general to say that joints open the way to weathering. In sedimentary rocks, the bedding-planes between successive layers of rock act in the same way. Plate 8 shows the effect of chemical weathering on bedded rocks, where salt spray from the sea is opening up the joint-planes and selectively attacking a series of thin beds by dissolving their limy cement. A section cut in rocks of any kind – for example, the face of a working quarry or a cliff on the shore – usually reveals the way in which weathering is guided by natural fissures. It also shows the downward succession from soil at the top, through well-weathered rock beneath, to rock which is but slightly loosened, and finally to rock which is still sound (Plates 1, 9).

Although complex chemical reactions may be involved, it can be said that the solid end-products of rock-rotting outside the tropics are sand and clay. The sand consists of chiefly grains of quartz, the crystalline form of silica ($SiO_2$), while the clay is composed of microscopic flakes of the clay minerals. These are hydrated silicates of alumina – compounds of aluminium, oxygen, silicon, and water, such as kaolinite ($Al_2O_3.2SiO_2.2H_2O$). Quartz and the clay minerals are chemically stable in the climates of most places outside the tropics, and continue to resist weathering long after other substances have been dissolved away.

Complex effects produced in the humid tropics include the instability of quartz and the formation of kaolinitic clays. Clays are diffused downward. Silica passes into the rivers, where it is more abundant than in most midlatitude streams. Otherwise, however, large volumes of discharge keep concentrations of dissolved materials down at low levels. The well-weathered waste-mantle of the humid tropics is composed mainly of compounds of aluminium, iron, oxygen, and water. In areas of seasonal rain, the dry season can bake the surface hard, while the wet season turns it into sticky red mud. Permanent drying-out, caused by forest clearance, down-cutting of rivers, change of climate, or any combination of these, results in the formation of a permanent, thick, cemented crust.

Many tropical soils would be barren by midlatitude standards. They lack soluble matter, lime, potash, phosphate, nitrogen, and humus, except for what is recycled within the forest. Weathering can penetrate 300 feet below level ground, opening the way to the downward movement of abundant water, and promoting the removal of all soluble materials. Despite their exuberant vegetation, tropical forests live on poor soils.

Wherever it occurs, and whatever form it takes, weathering prepares broken and rotted rock for movement downhill. Downhill movement has a single cause – gravity – but takes a number of distinctive forms. On some slopes, the waste-mantle moves so slowly that its motion passes unnoticed. Elsewhere, tons of loose rock suddenly hurtle down. All the movements contribute to the removal of weathered rock and to the denudation of the lands.

Movements of rock-waste down open slopes are distinguished from the transport of solid material by rivers. It is found, in practice, that rivers commonly discharge a far higher volume of water than of sand or mud. The Mississippi, renowned as being too thin to cultivate but too thick to navigate, is recognizably a river. There is a fairly sharp transition from a very muddy river to a very wet mud-flow, although it would be interesting to know how muddy a river would have to be before it would be classed as a stream of mud instead of a stream of water. One signal difference between rivers and mud-flows is that the latter are always intermittent, while – in wet climates at least – most rivers never cease to run.

Mud-flows stand at one end of a series of movements of the waste-mantle, with falls of dry rock at the other end. From time to time, sudden movements of rock-waste are reported in the daily press, usually under the name of landslides. There are, however, several kinds of landslide. Cliffs undercut by the sea, and steep banks undercut by rivers, are liable to collapse. If the rocks are strong and coherent, the undercutting goes on until, all of a sudden, great masses crash down. In clayey rocks another type of movement occurs, affecting unweathered rocks as well as the waste-mantle. This is the movement of slumping (Plate 10). It involves shearing of the rocks and rotation of the slumped mass. A scar is left where the mass has torn away, and a jumbled mass of debris marks the foot of the slide.

Slumping is widespread and frequent on steep slopes cut in clay, for clay acts as a colloid in taking up large quantities of water. In this way the strength of the clay is reduced, and shearing occurs. Failures of railway embankments and of railway cuttings after heavy rain are usually due to slumping. A long line of slump can be seen on the coast of north Kent, east of Herne Bay (Fig. 1), where London Clay is exposed in cliffs above the sandy Oldhaven Beds. When the London Clay is well wetted by the rain, soggy masses shear away, pouring over the steep bases of the cliffs and spreading in muddy lobes over the beach.

Flow at the front of the moving mass is promoted by a kind of auto-lubrication. Clay subjected to sudden shock can release the moisture trapped in its tiny pore-spaces, changing immediately into highly mobile slurry. This is why slumping on the sides of cuttings is often accompanied by flows of mud across railway tracks. Slurrying also occurs, in rather peculiar conditions, in the valley of the river Göta in southern Sweden.

The lower valley of the Göta is partly filled with silt, the uppermost 3 feet or so having dried out as a coherent cake (Fig. 2). Since the land has not long been relieved of the weight of an ice-cap (Chap. 13), it is gradually rising, and the river Göta is cutting

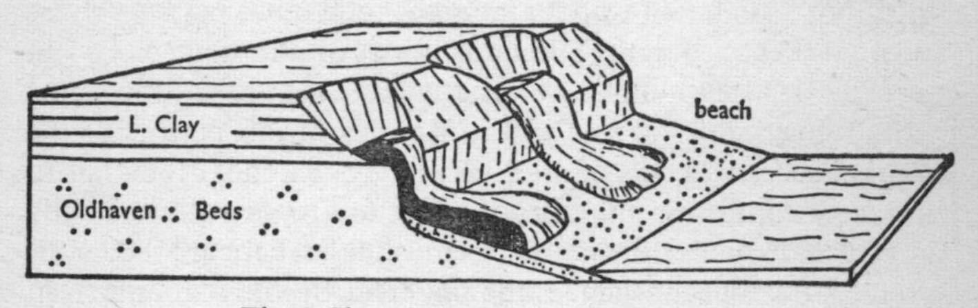


Fig. 1 Slumps on the north Kent Coast

steadily down into the silty fill. Every now and then a raft of dried silt breaks away and slides towards the river. The adjacent dry silt, left without support, also cracks off and starts to move. Repeated breaks can extend the slide right across the valley-bottom. Meanwhile, the shocks of the movement cause the moist silt beneath the moving rafts to slurry, so that the dry cakes travel forward over a layer of liquid mud. Slides which occurred in 1953 were triggered off by the vibration of a passing train, but it is known that intermittent sliding has been going on for many centuries; old slides appear as projections of the river-banks.

These slides are exceptional, in that they occur on gentle slopes. Slides on steep slopes are far easier to comprehend. The waste-mantle becomes thicker as weathering works deeper and deeper into the underlying rocks, and as rock-fragments falling from above are temporarily lodged. In glaciated valleys there is often a plaster of boulder clay reaching high up the sides, which becomes unstable when it is made very wet. In some places, avalanches of debris roar down long, narrow tracks, excavating trenches which look like embanked gullies. As a rule, sliding follows unusually heavy rain, which increases the weight of the poised rock-waste and lubricates it at the same time.

Frequent slides of this type occur in Glengesh, one of the ice-gouged valleys of Donegal (Plate 18). The upper sides of the trough are channelled by the tracks of old slides, while the lower slopes are bestrewn with fans of fallen debris. Here, as in mountainous parts of Wales and Scotland, the slides can block roads with jumbled heaps of mud and rock. Although the damage and obstruction

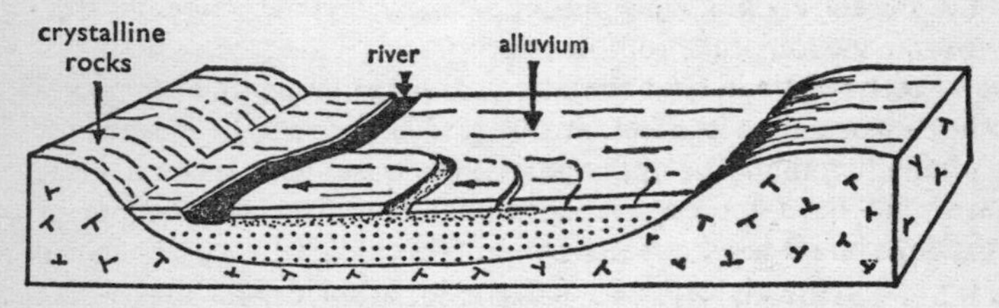


Fig. 2 Slides in the Göta valley, Sweden

caused by any one slide are often exaggerated in the distant press, it is possible for cars to be buried, bridges to be destroyed, and even for houses to be swept away.

The photograph reproduced in Plate 18 was taken on the day after sliding had occurred in Glengesh in the rainy August of 1957. In the August of the previous year, a number of similar slides were reported in the Cairngorms, where 5 inches of rain had fallen in a single storm at the end of July, and 4½ inches fell in three days in mid-August. Some of the Cairngorm slides cut channels 40 feet deep in the underlying boulder clay. In May of 1953, the Fort William district of Scotland was affected by slides occurring after 3 inches of rain had fallen on the high ground in 7 hours. One of the avalanches swept with it a boulder weighing 8 tons. More than 1000 tons of fallen rock had to be shifted from the road on the north side of Loch Leven, where three huge fans formed at the mountain-foot.

Rapid chemical weathering in equatorial conditions maintains a deep waste-mantle, but the root-mat of equatorial forests is typically shallow, because the trees do not need to go deep in search of water. Thus it is possible for mud-flows to occur beneath a cover of forest, without necessarily uprooting, overturning, or greatly disturbing the trees.

Landslides and mud-flows are sudden, dramatic, noisy, and localized. The type of movement known as *creep* is slow, unspectacular, silent, and widespread in its occurrence. Nevertheless its effects are great. Creep moves rock-waste down gentle slopes as well as down steep ones, and keeps the soil on steep slopes shallow in depth, even beneath a binding mat of roots. Wetting and drying, heating and cooling, thawing and freezing, and the action of earthworms and roots, all promote creep.

Creep occurs in cold climates, under the influence of temperature-changes, but its effects are likely to be obscured by movements of another kind. The upper part of the waste-mantle can become very mobile in summer, if the seasonal thaw does not penetrate far and the under-layer remains frozen. This topic will not be pursued at present, since it reappears in Chap. 15. In hot, dry climates, where rainfall occurs very seldom, debris which is too coarse to be shifted by the wind can form enormous sheets of creeping waste, with individual movements once again due to changes in temperature.

On steep slopes, the effects of creep are readily made out. Soil piles up against hedges, banks, and walls which run across the slope, and moves away from their downhill sides. Trees, tilted over by creep, compensate by curving their trunks at the base. On very many steep slopes a kind of ribbed pattern appears on the surface of the creeping waste, with little steps a foot or two in height running horizontally. These steps are called terracettes (Plate 11). An alternative name is sheep-tracks, but this title is grossly misleading. Terracettes can be found where no sheep have ever been. Moreover, although mountain sheep naturally make use of terracettes in moving across the slope, authentic sheep-tracks which run obliquely upslope do not take the form of steps. One of the clearest demonstrations that the formation of terracettes is independent of the treading of sheep comes from the Yorkshire Dales, where the steps run to the very feet of drystone walls and resume again on the far sides. No sheep could be malevolent enough to mislead a human observer by producing such an arrangement of treads and risers.

Little or nothing is known of the mechanism of terracettes.

Their formation does not appear to depend on the presence of a root-mat, for terracettes can occur on slopes with very little vegetation. Fig. 3 is drawn from a photograph taken in a Pennine valley, showing terracettes on an almost bare slope cut in unconsolidated boulder clay. There seems to be some connection between the angle of slope and the presence or absence of the steps, for in this example terracettes are confined to the steep bank which is being undercut by the stream.

On gentle slopes the action of creep is far less evident that on steep slopes, but its effects can nevertheless be detected. The waste-mantle, creeping across a valley-floor, can make river-banks overhang, quite independently of any undercutting by the river. In extreme cases, a little stream can be almost bridged. In the valley-heads of Epping Forest, where small streams are cutting into

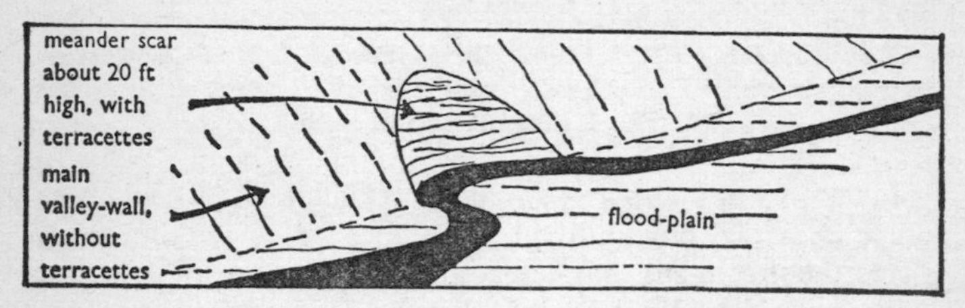


Fig. 3 Terracettes in a Pennine valley

London Clay, alternate wetting and drying cause the clay to swell and shrink by turn, shrinking mainly in the downhill direction and expanding downhill when it is again wetted. In many places it can be clearly seen that the creeping clay has moved over the pebbles of the stream-beds, binding them firmly down.

It is difficult to separate the effects of creep from those of surface-wash. The foundations of two Roman villas in England have been excavated from beneath 8 feet of rock-waste, but part of the accumulation is certainly due to rain-wash over ploughland above the villa sites. Assuming that the villas were abandoned in A.D. 450, the 8 feet of deposition represent about 1 foot for every 200 years, with creep contributing an unknown fraction to the whole.

Direct measurement of the speed of creep can be made by several methods. The simplest is that of opening a trench downslope, inserting a line of vertical markers, covering the marked face of the trench with waxed paper, and filling in the excavation. The cut can be reopened from time to time, and the movement of the markers recorded (Fig. 4). Typically, the rate of creep is greatest at a level slightly below the ground surface, decreasing downward thereafter. When the varying speed at various depths is integrated, a representative speed of about 1mm/year appears, for the total of the creeping mantle. In areas of water-absorbent clays, however, and on steep hillsides, creep can act 40 or 50 times as fast.

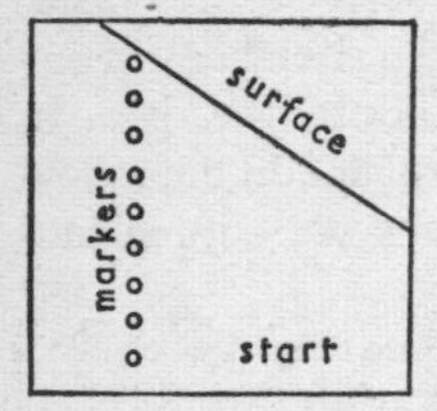

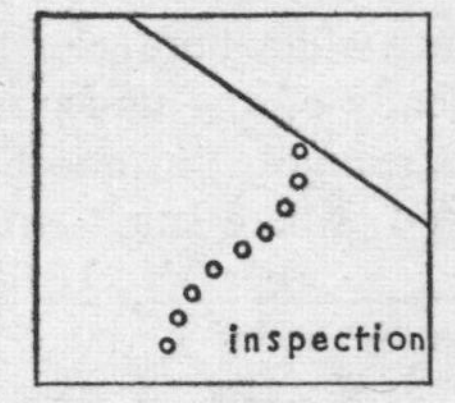

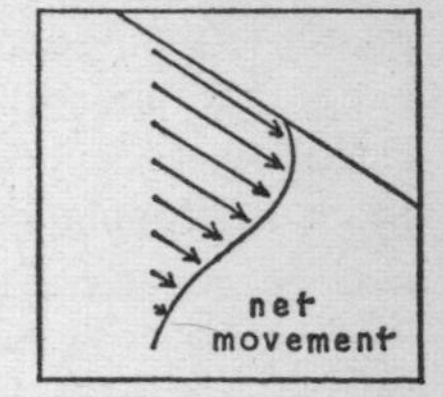


Fig. 4 Measurement of creep

In one way or another – by creep, rockfall, or landslide – rock-waste is carried downhill. In landscapes which are drained by rivers, the stream-cut channels control the evolution of the valley-walls, and it is the deepening of valleys which sets off movements of the waste-mantle. We might, therefore, have begun this chapter with the cutting of valleys rather than with the weathering of rock. But, as will shortly appear, the topic of rivers and river-valleys leads in a different direction altogether.

## CHAPTER THREE

# *River-Patterns and Landscape-Patterns – I*

CERTAIN types of soil-erosion clearly illustrate the transition from wash across a broad surface to flow in well-defined channels. As generally employed, the term *soil-erosion* implies the removal of soil from farmland, either by running water or by the wind. It is classified under three heads – blowing, sheet-erosion, and gullying. Blowing – i.e. erosion by wind – will not be discussed for the present. Sheet-erosion and gullying, which are the work of running water, occur in natural conditions as well as on farmland. But where farmland is affected, both of these processes act with devastating speed, providing speeded-up experiments which illuminate the study of erosion in wholly natural circumstances.

In sheet-erosion, water running across the whole surface strips away the topsoil. It is a slow process in the climatic conditions of western Europe, mainly because west European soils have structures which make them resistant to surface-wash. But it can occur when heavy rain falls on ploughed land – it was reported from parts of England in April 1951 and in November 1957.

In many parts of the U.S.A. it can affect very gentle slopes. Its first effects are often imperceptible, but as the soil is gradually thinned down, the land becomes less and less productive. Each heavy shower carries away some of the remaining soil in sheets of muddy water. The flow becomes more and more violent as less and less soil remains to check the running water. Sooner or later, if nothing is done to prevent it, the runoff is concentrated into rills. Thesc, digging deeper and growing wider, grow into gullies (Plate 13). Once gullies have been formed, they extend themselves headward and throw out tributaries. The original gentle slopes are destroyed. Definite channels collect all the surface water. General washing has been superseded by stream-flow.

Farmland under attack by sheet-wash or by gullying shows how delicate is the balance between erosive and protective forces. Soil-erosion has followed the plough, laying waste the land where the natural turf has been broken or the natural forest cleared. But it would be unjust to blame unwise farmers for all the devastation. Savage destruction of land in the interior of the U.S.A. and in New Zealand had already begun before the first settlers came. Settlement happened to coincide with a new phase of natural erosion. Analyses of rainfall records in the U.S.A. show that the change from very slow to very rapid erosion was caused primarily by a change in the heaviness and frequency of rain-showers.

One can easily imagine that rain falling on a newly uplifted land-surface would not take long to form channelled streams. For the purpose of discussion, it will be assumed that there is enough rain to sustain permanent rivers – that is, that some water is left over after the soil has been charged with moisture, and after allowance

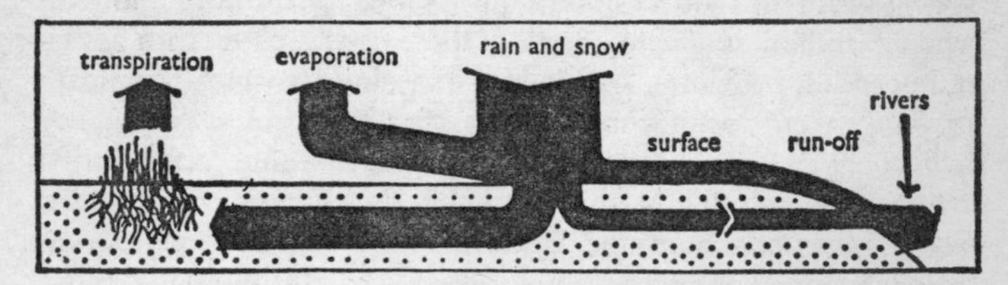


Fig. 5 Precipitation and runoff

has been made for losses by direct evaporation and for transpiration by plants. The total loss to be expected is surprisingly high – in southern Britain, where some 30 inches of rain fall in the average year, the surface-runoff is equivalent to about 10 inches. Two-thirds of the total fall is lost. It is the remaining one-third which fills lakes, feeds springs, and nourishes permanent rivers (Fig. 5).

Rivers control, in large measure, the development of the lands which they traverse. Rocks on the sides of their valleys are exposed to weathering and to the downward pull of gravity. Rivers carry rock-waste to the sea, either in solid form or in solution. Rivers, surface-wash, and the downhill movement of solid rock combine to remove the substance of the land. Given enough time, they can reduce the highest and most extensive landmass to heights little above the level of the sea. In the long span of time between the uplift which forms new land and the final levelling, distinctive patterns emerge, both in the arrangement of high and low ground and in the river-systems.

Few mountains have been built into shape. Most of them are the products of earth-sculpture. Volcanoes form an exceptional class, for they have been constructed on the spot. But volcanoes amount to no more than a small fraction of the world's total of mountains. Mountains of other kinds rise above their surroundings merely because surrounding areas have been lowered by erosion.

This statement needs qualifying, to the extent that mountains owe their height to uplift. But there is an important distinction between the general uplift of a mountainous area and the individual rise of single mountains. In the early days of geology, when little was known about movements of the earth's crust, it was possible to believe that mountains had been formed by localized upheaval. When it was seen that local uplift is the exception and regional uplift the rule, attention was paid to the complementary relationship between mountains and valleys. At one time it was supposed that valleys originated as cracks in the crust – cracks produced during upheaval of the land. This slightly more reasonable, but still largely mistaken, idea was based on a very short estimate of geological time.

When the whole surface-history of the earth was crammed into a few thousand years, it was natural to suppose that a single catastrophic episode of crustal movements had taken place. But as the available time was extended to lengths of million of years – indeed, to thousands of millions – it became clear that folding, breaking, and erosion of rocks could be explained by slow processes of change. Bit by bit, proof accumulated that the crust is still unstable. Appeal to some past catastrophe, in order to account for observed facts, became unnecessary.

Mountains and valleys were recognized as the products of selective erosion. It seemed natural enough that valleys should be located on weak rocks, while rising ground should be based on strong rocks – facts which inevitably impressed the first makers of geological maps, who were able to follow lines of hills in tracing resistant formations across country.

If this were all, the scientific study of landscape would be a simple matter. In practice, however, it rarely happens that the pattern of drainage faithfully reflects the distribution of the underlying rocks. It can be scarcely imagined that an *original* stream-system will fit neatly on to the plan of rocks and structures. If, therefore, valleys are in fact cut exclusively in weak rocks, and hills based exclusively on strong rocks, the drainage is taken to have become adjusted to structure in the course of evolution. In actuality, partial adjustment is usually the most that can be expected in a drainage-basin of any size.

Five things, then, are taken into account when a stream-pattern is studied – the arrangement of streams, the distribution of rocks of different type, the geological structure, the relationship between streams on the one hand and geology on the other, and any sign of discordance. As will be seen, discordances between streams and structure give most helpful clues to the origins of drainage – it is the discordant reaches which preserve the lines of old stream-courses.

The first three examples to be described are meant to show close correspondence between the patterns of structure and of drainage. Three small areas have been chosen, in order to give simple illustrations of the effects on relief of variations in structure, and in order to exemplify principles which will be applied in later descriptions of more complex areas. No attempt will be made to classify patterns of drainage, for the success of such an attempt is controlled largely by the size of the area considered. Patterns which appear on a large-scale map of a small piece of country may well vanish when a larger area is represented on a smaller scale, and when the well-defined arrangement of minor streams is overshadowed by the notable discordance of main rivers.

In the Wenlock Edge country, long valleys have been cut in weak rocks while strong limestones stand up in bold relief (Fig. 6). The

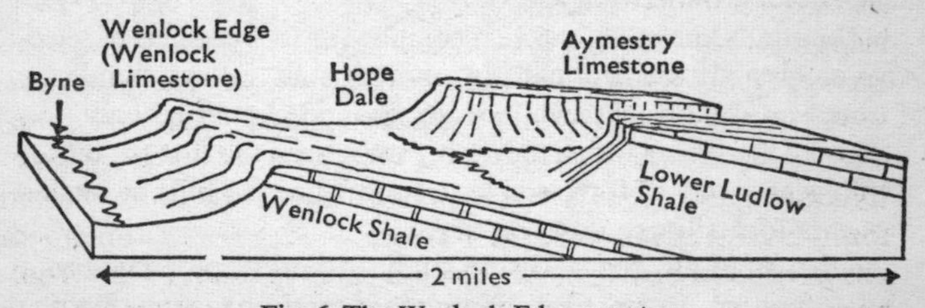


Fig. 6 The Wenlock Edge country

structural arrangement is one of the simplest possible – alternating formations of limestone and shale dip gently away in a single direction. Streams flow mainly along the valleys excavated in the shales, and the pattern of the terrain closely reflects the distribution, attitude, and relative strength of the rocks.

A complication is introduced when the rocks have been domed instead of merely tilted. A convenient small dome with which to begin is that of Hope Mansel, at the north end of the Forest of Dean (Fig. 7). In the core of the dome are exposed resistant sandstones. The flanks are covered by less resistant marly rocks, which have been stripped off the resistant core. Around the central hill two circular valleys have been cut in the weak marls. They are drained by streams which unite before escaping through a gap to the west. Finally, the outer walls of the valley are based on the strong rocks which overlie the marls.

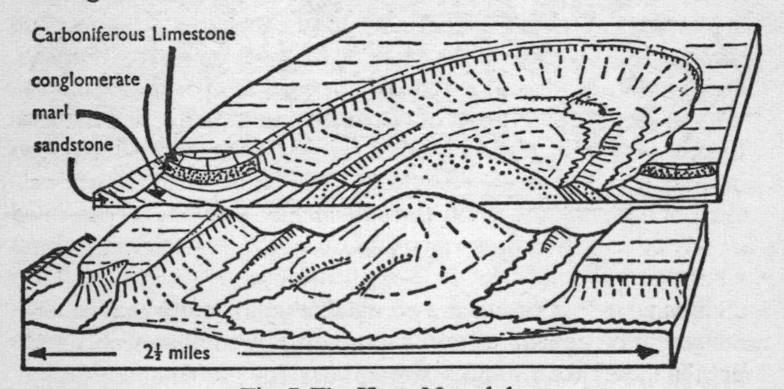


Fig. 7 The Hope Mansel dome

Domes with thick and varied covers are characteristically encircled by several concentric valleys, which are separated by prominent lines of in-facing hills. Such is the case with the Woolhope dome (Fig. 8), where a resistant core of Llandovery Sandstone and Woolhope Limestone is encircled by Wenlock Shale (weak), Wenlock Limestone (strong), Lower Ludlow Shale (weak), and Aymestry Limestone (strong). Rings of valley have been cut in the two formations of shale, while the Wenlock and Aymestry Limestones rise in rings of high ground. The landscape-pattern of the Woolhope dome may aptly be contrasted with that of the Wenlock Edge country since both are based on rocks of similar kind and similar age.

Each of these three localities displays the effects of differential erosion. Their limestones and sandstones are strong because they are bound together with natural cements. In addition, they are to some extent permeable, allowing water to soak through them. They are in consequence partly immune to the influence of surface-wash. The shales, on the other hand, are uncemented and impermeable. As well as being mechanically weak, they prevent water from soaking through the surface. Hence they are readily eroded.

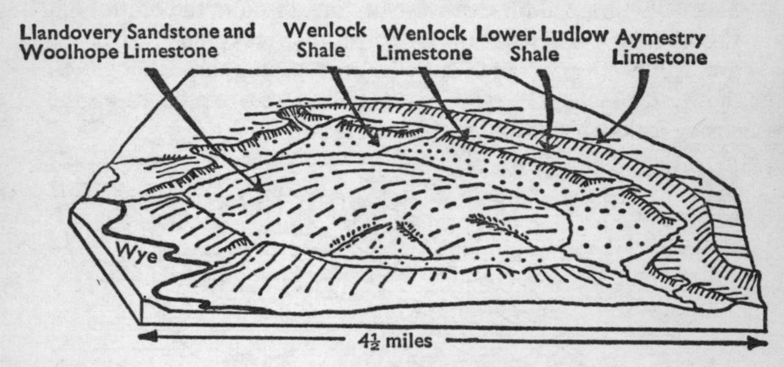


Fig. 8 The Woolhope dome

It is usual to find strong and permeable sedimentary rocks alternating with other sedimentary rocks which are impermeable and weak. In these circumstances, streams develop preferentially along the lines of weak impermeable rocks. It is in this way that adjustment involves the growth of some streams at the expense of others. All streams are engaged in ceaseless competition for territory. Weak rivers are robbed of tributaries, lose parts of their basins, and can be entirely dismembered. Strong rivers extend their basins, capture parts of the weak competitors, and become all the more powerful as they succeed.

River-capture – also known by the graphic name of river-piracy – is most easily explained by reference to the type of structure illustrated by Wenlock Edge – i.e. uniclinal structure, in which sedimentary rocks dip steadily away in one direction. The dip is not a feature of deposition, but is due to tilting. Tilting of this kind has often been produced by an uplift which, raising a heavily denuded crustal block, also warped the sediments laid down against its flanks. Since we are dealing with river-erosion, we assume that part of the sedimentary series is raised above the sea. Rivers form on the emergent land, running downslope in the same direction as the dip of the rocks.

It is reasonable to suppose that the land traversed by these new rivers is underlain by resistant material, if only because weak sediments would probably be removed by wave action as the land rose. As soon as one of the rivers breaks through the resistant cap, the alteration of the drainage-pattern begins.

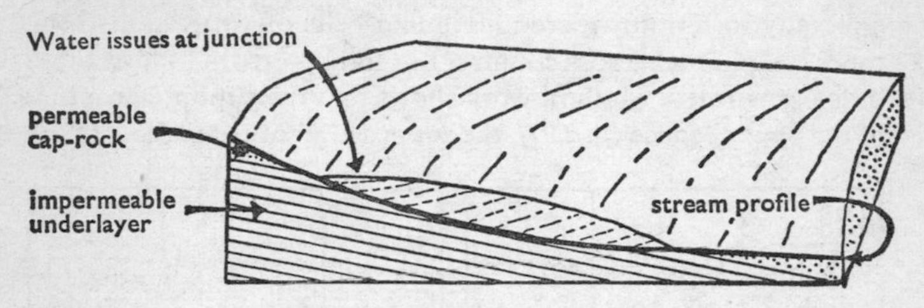


Fig. 9 Breaching of a cap-rock

The cap is certain to be broken through, for rivers deepen their valleys more in the middle courses than near the heads or near the sea (p. 80). When a breach is made, water percolating through the resistant cap seeps out at the junction with the underlying clay (Fig. 9). Seepages become springs, and springs work back into the valley-side. The valley-walls are cut back, and little streams connect the receding springs with the main river (Fig. 10).

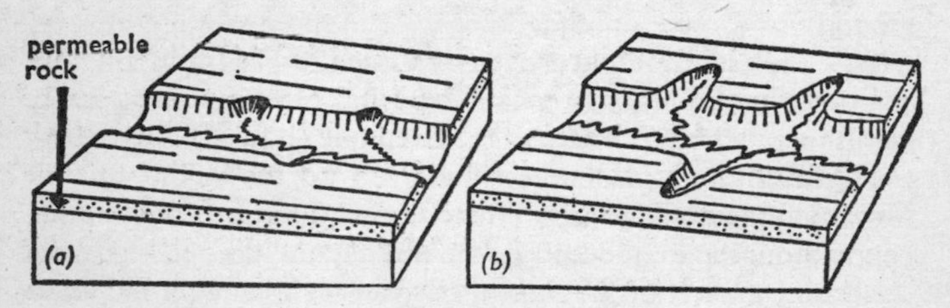


Fig. 10 Receding springs

In the course of time these little streams become large tributaries. The best favoured of the original streams throws out the most powerful tributaries, which press back the watersheds into the territories of their competitors. Capture takes place when the head of a tributary works right back to the line of another river. The latter is diverted (Fig. 11).

Past captures are characteristically recorded by sharp bends – elbows of capture – in river-courses. Where such bends occur opposite dry gaps in high ground, there is good reason to suppose that capture has taken place – at least, to make the supposition for the sake of argument. For some time it has been a common practice to reconstruct dismembered rivers with the aid of maps and rulers, especially in terrain where alternating belts of weak and strong rocks run parallel across country. But serious errors have at times been perpetrated. Modern work shows that the supposed capture of the upper Meuse (p. 219), the supposed piratical success of the

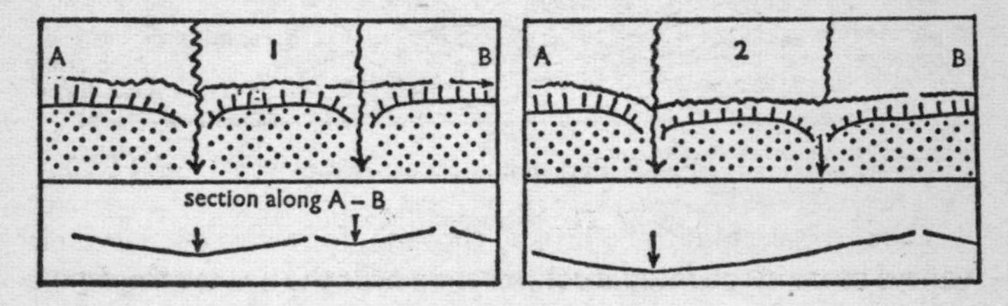


**Fig. 11 River-capture on tilted rock**

Spey (p. 178), and the whole history of the Warwickshire Avon (p. 227) must be re-interpreted. Research is coming to rely not on straight lines drawn on small-scale maps but on a study of the ground.

The readiest proof that a given river once flowed through a now-dry gap is supplied by river-laid deposits. This statement can be fitly illustrated by reference to the Evenlode (Fig. 12). Terraces of river-gravel alongside the Evenlode run up the valley into the Moreton Gap, disappearing where the head-streams of the competing Stour have cut deeply down. But beyond the head-valley of the Stour, gravels of the same series can be located in the notch between the Cotswolds and the outlying Ebrington Hill. As shown in the diagram, the two sets of gravels lie on a single curve – the old profile of the Evenlode, developed when the river rose beyond the present watershed, and before it had been beheaded by the vigorous Stour.

The Hope Mansel Dome, the Woolhope Dome, and the Wenlock Edge country have had far more complicated histories than that outlined above for a hypothetical region of gently tilted sediments. Few parts of the world's lands are, in actuality, based on rocks which have undergone a single upheaval. Nearly all the upland regions which have been studied in detail show signs of repeated uplift, and portions of old erosional plains survive on their summits. Nevertheless, the three examples which have been

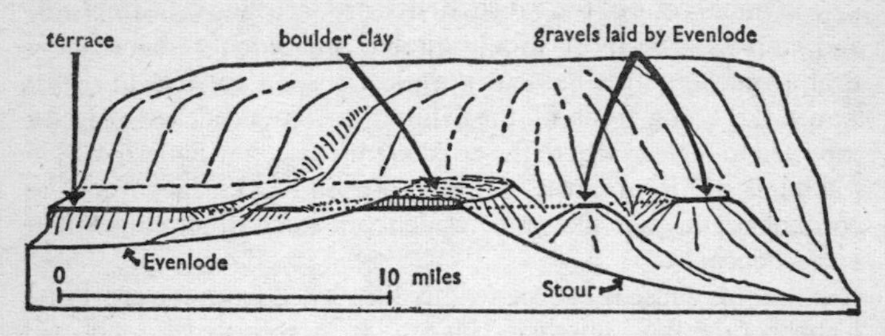


Fig. 12 The beheaded Evenlode

quoted illustrate the way in which streams flowing down the dip of sedimentary rocks are likely to be dismembered, and superseded, by streams developing at right angles to the dip.

Even in regions of very simple structure, however, drainage is not always well adjusted. Nor is a structural uplift always reflected in rising ground. A well-known case in point is that of the Nashville Dome in Tennessee, where the central area has been excavated to form low ground.

Although the Nashville Dome has undergone prolonged upheaval, it appears on the relief map as a depression, 60 miles wide and 120 miles long, and lying as much as 600 feet below the level of its rim of hills. Doming began here about 300 million years ago, and has been intermittently renewed in subsequent times. The cover of strong cherty limestones has been stripped off. Wide valleys have been cut in the underlying rocks, which are both thick and weak. There is no resistant formation in the core capable of forming a central block of hills. The ground is lowest precisely where uplift has been most marked.

Those rivers which cross the Nashville Dome – the Cumberland and its tributaries – were presumably initiated on a gentle erosional surface which was warped down towards the west. Rivers were already in being before up-doming began. They have maintained themselves across the rising structures, having been able to cut downwards as fast as the crust rose. Rivers of this sort, which are not diverted by structures which grow athwart their course, are said to be *antecedent* – that is, antecedent to the uplift.

Antecedent rivers are by definition discordant to structure, and antecedence is one of the causes of discordance between structures and streams. The most widely quoted antecedent stream is the Brahmaputra, which has cut gorges thousands of feet in depth across the rising folds of the Himalayas. Antecedence is to be expected in regions where the crust is unstable, particularly where folding is still in progress. On rigid crustal blocks, however, discordant drainage is likely to result from some other cause than antecedence.

What this cause is can be readily illustrated from the erosional history of the Lake District. This area is another structural dome, but in direct contrast with the Nashville Dome the Lake District Dome has a very resistant core. Furthermore, it displays a pronounced radial pattern of drainage (Fig. 13a). Its radiating valleys have been emphasized by deep glacial grooving during the Ice Age, when ice-tongues gouged out the long troughs of the lake-basins.

Ancient sedimentary and volcanic rocks form the core of the Lake District. Involved in an episode of mountain-building, these rocks were intensely crumpled. Their structures are complex and disordered. Crumpling and uplift were followed by severe and prolonged erosion, which succeeded in reducing the land to low levels. Submergence followed, with the denuded land sinking gently beneath the sea. On the sea-floor were laid down new rocks beginning with a thick layer of Carboniferous Limestone. Since the limestone beds were originally flat, their structure is greatly at variance with the disordered structures of the older rocks beneath – that is, the Limestone is unconformable on the older rocks, and

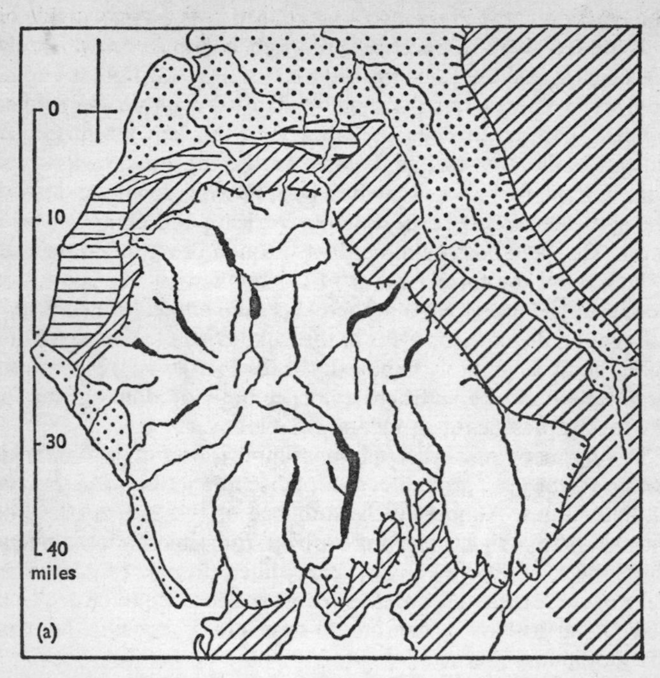


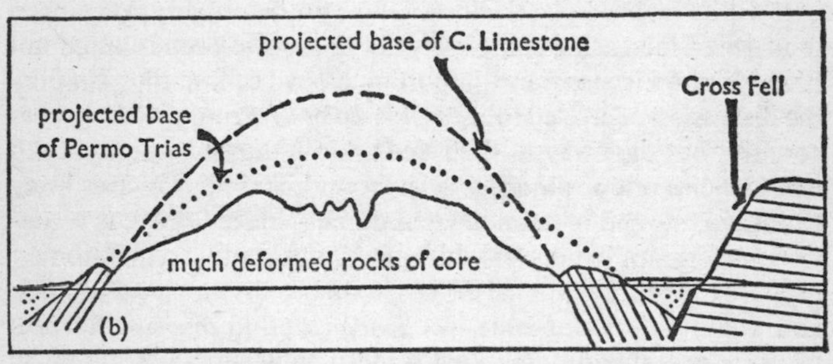


The section is drawn on the same horizontal scale as the map; vertical exaggeration of section is x 12.

Fig. 13 The Lake District dome. Key to shading: stipple, Permo-Trias; diagonal ruling, Carboniferous Limestone; blank, rocks of core

the junction between the two series constitutes an unconformity.

Renewed earth-movements produced a dome, on which the Carboniferous Limestone formed a cover (Fig. 13b). As the dome rose above the sea, rivers coursing radially down the sides formed upon it. Cutting through the cover, and through the surface of unconformity, they incised themselves into the tough rocks of the core. On the Limestone they had been flowing down the dip. On the core they ran regardless across intricate and complex structures. They had become discordant through being superimposed.

Nearly all the former cover of Carboniferous Limestone has been stripped away. No more remains than a low, broken belt of hills round the outer margins of the Lake District. But even if the cover had been entirely removed, the distinctly radial plan of the drainage would be sufficient evidence both of doming and of the former presence of an unconformable cover.

In very many upland areas, superimposition rather than antecedence is the most probable cause of discordance between streams and structures. Almost all the drainage of the British Isles, for instance, shows signs of having evolved from superimposed rivers. But because of selective erosion along lines of weakness and along belts of weak rock, discordant streams are doomed to be dismembered – particularly by capture. In some areas, very little remains of the original pattern.

It will be realized that the localities so far described are arranged in order of increasing complexity. Domes have been selected not because of their structural importance but because they simplify the discussion of eroded folds. Folds do not run indefinitely across country, but die away at their ends. Now the end of an upfold, where it dies out or plunges underground, is similar in every way to the narrow end of an elongated dome – indeed, there is a continuous series of structures which, beginning with circular domes, passes on through elliptical domes, to markedly elongated domes, and finally to long upfolds. An eroded upfold displays features of the kinds already described – steep infacing edges based on strong rocks, long valleys carved in weak rocks, and high or low ground in the middle according to the strength or weakness of the core.

Lines of hills converge at the end of a fold in a smooth nose, as is well shown by the eroded fold of Plantaurel in the Pyrenees (Fig. 14). Along the flanks of this fold – about 6 times as long as it is broad – infacing hills rise in parallel belts, reaching summits about 2500 feet high. The even crests belong to some past period when a level surface was eroded across the folded rocks. It is from this surface that the rivers have been superimposed across the axis of the fold, so that the two trunk streams cut right across from one side to the other.

In regions of dry climate, where rocks lie bare to the sky, fold-structures can be revealed in full clarity (Plate 79). Weathering fails to soften the contours of the infacing edges, which frequently

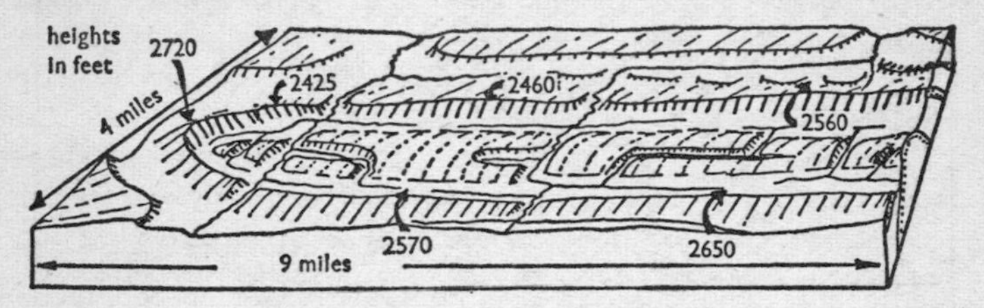


**Fig. 14 The long upfold of Plantaurel**

take on a jagged outline like that of the dog-tooth ornamentation of Norman architecture. The triangular hill-masses are called *flatirons.*

Whereas the map of a small part of a folded belt is likely to reveal a trellis-like pattern of streams, the map of a larger area shows the convergence of valleys and of streams towards the noses of the individual folds. A folded region poses a far more complex problem than does a single fold, for the origin of the regional drainage needs to be explained. There are usually two possibilities – either that the original streams ran, as directly as possible, down the two flanks of the folded belt, or that they ran right across it. In the former case, uplift was greatest in the centre; in the second, tilting occurred on a regional scale.

It is important to distinguish between the compression responsible for folding and the general uplift of the folded region. Geological research shows quite clearly that intense folding characteristically precedes general uplift. Furthermore, tens of millions of years may be required to complete the folding, and general upheaval is typically slow at first. Thus there is no reason to think that when a tightly folded belt begins to rise its surface is bound to be corrugated. The rivers which form on the rising mass can flow down the flanks or across the region from one side to the other, according to the nature of the uplift.

One of the best-studied regions of tight folding is the Appalachian region of the U.S.A. (Fig. 15). Very many of its valleys correspond to belts of weak rock. Strong formations underlie high,

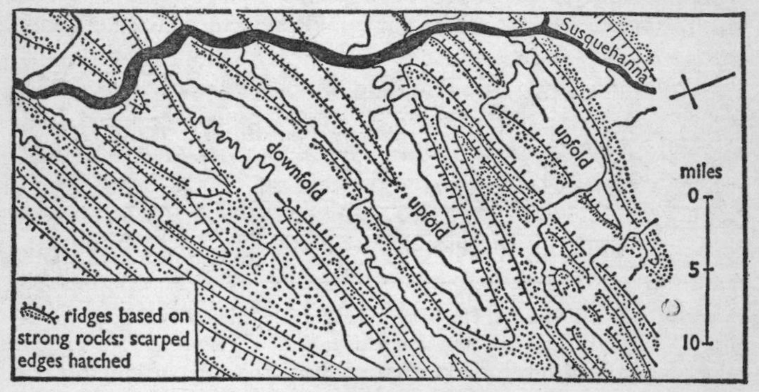


Fig. 15 Appalachian folds

long, and narrow belts of hills, which converge towards the noses of dying folds. But although the minor streams are well adjusted to structure, the major rivers break across the pattern of folding. In places, admittedly, some of the main rivers are manifestly structure-guided, but in general they are markedly discordant. They have obviously been superimposed on the structural pattern of the region.

Beyond this clear inference, much remains obscure. It is known that the Appalachian region, like the Nashville Dome, has been intermittently uplifted, and the complications due to repeated upheaval make it difficult to reconstruct the original pattern of rivers. So complex is the problem that several highly contrasted reconstructions have been suggested.

Evidence of uplift is supplied by erosional platforms at various levels. Each platform cuts broadly across the closely folded rocks, being preserved as horizontal crests on the narrow ridges. So even are the ridge-crests that they can only be explained by past episodes of widespread erosional levelling. But in so large a region, where the land-forms have not all been mapped in detail, it is not easy to correlate all the series of erosional flats with certainty. Consequently, varying opinions have been held about the nature and extent of the varying uplifts, and about the systems of rivers which ran over the uplifted surfaces.

At the outset, the problems were wholly geological. The thick sedimentary rocks of the folded Appalachians represent rock-waste stripped off some old land-surface. It used to be thought that the sediments came from a huge, high-standing crustal block which, lying east of the present Appalachians, has disappeared by foundering. Directly or indirectly, this hypothesis is responsible for the view that the original streams ran right across the Appalachian region from one side to the other. In one interpretation, the original drainage was dominated by streams which rose in the northwest and flowed to the southwest, initially down a tilted slope. A diametrically opposite scheme includes rivers which ran across the region from southeast to northwest, and which have been largely reversed by regional tilting – the tilt being presumably due to the foundering of the supposed crustal block to the east.

A third – and most likely – interpretation is that broad upwarping has repeatedly occurred, causing streams to flow down both sides of the highland belt. This view agrees with the known behaviour of other folded belts. It means that the present Appalachian drainage has evolved from streams flowing either towards the northwest or towards the southeast, which have modified the crestal watershed by competition and capture, and which have become adjusted, in places, to the structures of the Appalachian folds.

There seems to have been an essential difference in time between the formation of the tight folds and the last general upheavals of the Appalachian area. The main folding movements took place about 250 million years ago, deforming sediments which had accumulated in a long trough between two belts of mountains – not, as formerly thought, at the edge of a large stable block. Folding culminated in general upheaval, which exposed the Appalachian region to erosion. In the course of 100 million years, the fold-mountains were levelled. Subsequent history – the history of the last 100 million years or so – begins with submergence of disputed extent, which caused the eroded folds to be concealed by an unconformable cover, and continues with repeated broad upwarping during the last 50 or 75 million years. It was the first of the broad uplifts which raised the Appalachians in something like their present form, initiated rivers on the two flanks, and caused the unconformable cover to be attacked. The existing drainage has developed from streams which have been superimposed on the tight folds, either from the unconformable cover or from a very gently sloping erosional surface. Rock-waste stripped off the upland has gone to form sedimentary rocks, either in the Mississippi Basin or on the Atlantic Coastal Plain.

CHAPTER FOUR

# *River-Patterns and Landscape Patterns – II*

ALL the localities in the previous chapter include permeable as well as impermeable rocks. These rocks have been broken as well as bent – that is to say, faulting has occurred in addition to folding. But since, in each case, faults merely complicate the pattern of fold-structures, without dominating the structural pattern, it has been possible to ignore their effects in concentrating on the surface expression of folds. Similarly, it has been enough so far to note the contrast between permeable and impermeable formations, without giving special treatment to one or other of these classes. Under the present head it is proposed to deal with the terrains of regions wholly underlain by highly permeable rocks. The topic of faulting, and its expression in the landscape, will appear later (Chap. 5).

In a given climate – assuming it to be wet enough to sustain permanent streams – the texture of drainage is strongly affected by the extent to which the underlying rocks are permeable. Two extremes can be imagined. At one extreme, where the rocks are wholly impermeable, percolation is impossible. Runoff is confined to the surface. At the other extreme, the rocks are so highly permeable that there is no surface drainage whatever. Although it is rare indeed for a rainy region to be quite streamless, some extensive pieces of limestone country do, in fact, carry no surface rivers, and others possess very few.

Many limestone terrains display whole suites of land-forms of a most distinctive kind. For this reason, limestone country usually appears as a separate topic in works dealing with earth-sculpture. In regions underlain by rocks other than limestones, structure usually exerts a strong influence upon the pattern developed by streams, and consequently upon the whole pattern of the terrain. Besides, rock-types which pass under a single name can vary widely in their response to weathering and erosion. Some sandstones are poorly cemented, weak, and highly permeable, while others are well cemented, strong, and resistant to percolation.

Although rocks which have crystallized from a molten state are typically resistant, the land-forms developed on them raise problems of distribution rather than of rock-character. It is true that areas of plateau basalts are distinguished by columnar jointing, and often by canyons, while granite masses that have been deeply rotted disintegrate into the seemingly jumbled heaps of tors. However, regionally distributed crystalline rocks are predominantly those altered by major crustal movements. They occur in the antique continental cores of northern Canada, the northern fringes of the Baltic Sea, parts of Siberia, central Brazil, portions of Africa, and the western two-thirds of Australia. They are also known from the former mountainous margins of ancient continents, as in Norway and the Scottish Highlands. In all these areas, lines of weakness are likely to be lines of faulting, as opposed to belts of inherently weak rock; in some cases, as in Scotland and Norway, lines of faulting have been emphasized by glacial gouging.

Limestone country, then, provides the most satisfactory illustrations of the visible effects of rock-character upon landscape. In contrast to rocks of most other kinds, limestones as a class have one important property in common – solubility. It is because of this property that limestone country can be distinguished from other kinds of terrain, and can be made the subject of systematic description. Nevertheless, there are considerable variations within the class, and some limestones fail notably to display the full range of characteristic features. In order that anticlimax may be avoided, the following account begins with the least spectacular kind of limestone country – namely, that developed on Chalk.

Chalk is a remarkably pure limestone, composed almost wholly of calcium carbonate ($CaCO_3$). As limestones go, it is rather soft. It is too weak to support great weights, and rarely forms the roofs of large natural cavities underground. In this respect it contrasts strongly with some other limestones. But Chalk is well jointed, cut by fissures which can be widened by solution. Thus water can percolate freely into it – so freely that many Chalk areas are streamless, as can be well seen on the Chalk downland of England and on the Chalk plateaus of Normandy. Underground water is dispersed along the many planes of jointing and of bedding.

Despite its lack of surface-water, Chalk country is usually dissected by systems of dry valleys (Plate 12). It is conceded that these valleys have been cut by running water. The problem is to explain where the streams have gone.

In one view, the valleys were cut by streams which flowed over frozen subsoil during the coldest parts of the Ice Age. Permanent frost is supposed to have sealed the fissures of the Chalk with ice, preventing percolation and confining runoff to the surface. Although frozen ground is known to have been widespread in former times, the interpretation of dry Chalk valleys related to freezing at depth has lost favour, mainly through competition from another idea. In this second view, the valleys have gone dry in the natural course of erosion. It is suggested that, as valleys are cut in the weak rocks bordering the Chalk, the level of saturation within the Chalk is lowered, so that springs run dry and streams cease to flow (Fig. 16).

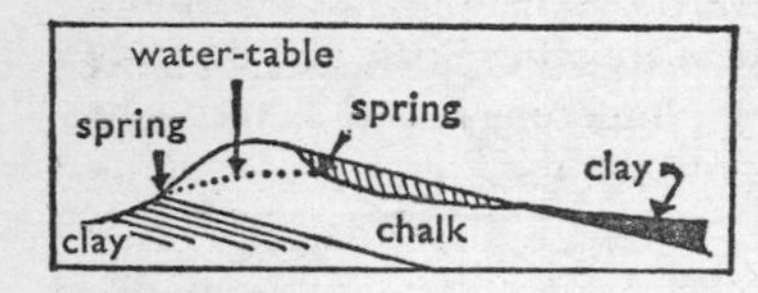


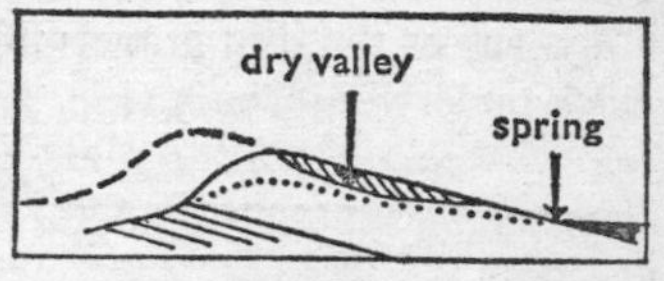


Fig. 16 A suggested origin of dry Chalk valleys

Against both of these attempted explanations must be set the results of recent work in the field. Valleys, now streamless, in the face of the Chiltern Chalk are known to have dried up during the post-glacial period (Plate 14). Since the time when they last carried surface streams there has been very little deepening of the main valleys, and very little excavation of the water-retaining clay beneath the Chalk has taken place. At the least, grave doubt is thrown on the idea that drying-out had been due to the lowering of the water-table by erosion. Moreover, the springs in these particular valleys failed far too late for their failure to be related to the thawing of frozen ground.

Now it is well known that occasional streams flow in Chalk country after unusually heavy rain. In the wet spring of 1951, after a wet autumn and a wet winter, many valleys on Salisbury Plain were occupied by unaccustomed rivers. It may be wondered, therefore, if the dry valleys can be referred to a time when the climate was rainer than it is today. On a later occasion (p. 222) it will be seen that there is every sign of high rainfall and runoff at some former time – rainfall and runoff quite high enough to make streams flow in many of the Chalk valleys which now stand dry. It seems unnecessary, in seeking an explanation of these valleys, to look beyond the climatic changes which are indicated by independent evidence.

If any one land-form is typical of limestone country, it is the sink – the closed depression which swallows surface water. The most satisfactory sinks are those down which whole streams vanish. In the Chalklands of England and France there is little chance for surface streams to run on to the Chalk, simply because this rock stands up in relief. Exceptions, however, are known, and localized sinks are recorded. At Water End, in Hertfordshire, the little Mimms Hall Brook runs down a valley cut in impermeable London Clay, across the thin gravels of the Reading Beds, and on to the thick underlying Chalk (Fig. 17). Numerous funnel-shaped sinks

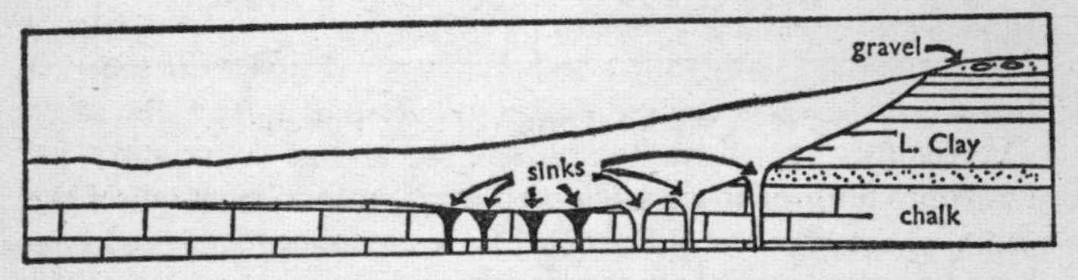


**Fig. 17 Sinks at Water End, Hertfordshire**

in the Chalk take the water underground, where it flows eastwards under the neighbouring hills to emerge in the Lea valley at powerful springs.

So many sinks have developed at Water End that, coalescing, they have merged into a single depression. Few are capable of working at a given time. Each is liable to become choked by mud and twigs, by the loose gravels of the Reading Beds, and by the old motor-tyres and empty oil-drums which accumulate in all streams near large towns. After heavy rain therefore, when the holes which happen to be open cannot take all the water which comes down the valley, a lake forms. If the lake overflows, its surface outlet carries water away to the northwest – not to the Lea, but to the Colne. The surface route to the northwest represents the course of the Mimms Hall Brook before the sinks were first opened.

For the sake of completion, it may be observed that an earlier valley, running northwards from the Water End site, has been plugged by glacial deposits. The briefest possible summary of local landscape-development must, in consequence, include the following items: (i) surface flow to the north, (ii) plugging of the original valley, and diversion of the surface-stream to the northwest, (iii) opening of sinks and partial diversion of the stream to drainage underground. One may guess that the Mimms Hall Brook is becoming less and less able to make a connection with the Colne. The depression formed by coalescent sinks is being continually deepened by the stream, and the reservoir which must be filled before the surface-outlet is used continues to grow larger. It may not be long before the course above ground is finally abandoned.

Exposures in numerous quarries (Plate 15) show that sinks in the Chalk narrow rapidly downwards. Holes 20 or more feet across at the surface taper off to shafts a foot or two in diameter. Some old shafts are known to go as deep as 60 feet. They lost their water to fissures opened along the planes of bedding, where cavities an inch or so deep are now packed with sand washed in from the pipes. Thousands of small hollows, about the size of active sinks, are known to pit the surface of bare Chalk country. Thousands more, filled with rock-waste and concealed beneath smooth ground, have been revealed in quarries. Most of them, like the sinks at Water End, have presumably been opened from the surface downwards. The usual view is that they are located at the intersections of master-joints, where dissolving water could most freely attack the rock. But sometimes a hole is formed by subsidence – that is, by the collapse of the roof of an underground cavity. Very little is known of the development in the Chalk of underground hollows, except that it seems to be rare.

Rather strangely, natural caves are still rarer in the limestone belt of the Cotswolds. Cotswold limestone is, mechanically speaking, far stronger than Chalk, but is no less soluble. Caves might well be expected, but seem to be absent altogether. Similarly, localized sinks might also be looked for in a region where the rocks are very well jointed, but they too are seldom encountered. The main difference between the Cotswold country and the Chalklands is that, whereas the latter is generally streamless and is gently modelled, the former supports numbers of surface streams and is trenched by steep-sided but rather widely placed valleys (Plate 16).

Steep valley-sides in the Cotswolds are developed on strong limestones – rocks coherent enough to be widely used for building. In the larger valleys, uneven walls can be seen of a type unknown in the Chalk belt; the steepest faces mark the edges of limestone formations, while gentler slopes are recorded on the shales which are interbedded with the limestones. It is the shales which throw out underground water as powerful springs. Where a stream crosses a belt of limestone and the water-table is well below the valley-floor,

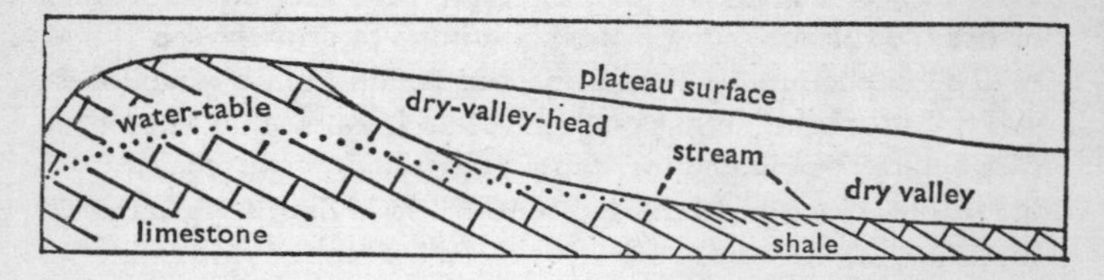


**Fig. 18 Discontinuous stream in the Cotswolds**

water is lost (Fig. 18). Hence many Cotswold streams are discontinuous. In some reaches they flow intermittently, according to the rise or fall of the water-table.

In most of the larger valleys, however, the underlying limestone is sealed off by thick alluvium, which prevents the water from soaking through the bottom of the stream-channel. Minor valleys and valley-heads in general, are streamless – at least for most of the year – and some of them show no trace whatever of stream-channels.

On limestones which are still more coherent, more massive, and stronger than those of the Cotswolds, additional features are to be observed. In the British Isles, the typical land-forms of limestone country are best and most fully developed on the Carboniferous Limestone, which occurs in thick, well-cemented beds. Dry valleys are extremely common on this rock (Plate 17). Sinks are numbered in hundreds. Some of them swallow quite large streams (Plate 19). Drainage passes underground to ramifying systems of caves and galleries, which direct it to numerous springs. These, the complement of sinks, are called *risings*.

On the surface, the most evident characteristic of Carboniferous Limestone country is bare rock. Soil disappears down slits, where joints in the massive beds have been opened by solution. In detail, joint-blocks are often seen to be furrowed by little channels where rainwater has etched the soluble rock (Plate 6). Vertical rock-walls appear where particularly strong beds are cut across. In some conditions a wall may even overhang (Plate 20).

This photograph referred to is of Gordale Scar, in the Craven district of Yorkshire. At this point, Gordale Beck slashes the edge of a limestone block in a narrow gorge. The features displayed serve to draw attention to a popular misconception. Seen from below, Gordale Scar is readily taken for a collapsed cavern, and is often so described. But when the stream is followed down from above, it can readily be seen that the gorge is nothing more than a surface-valley cut through the steep edge of a plateau. In all probability, the side-walls have been corroded at the base by spray flying from the cascading stream.

Gordale Beck flows over the surface because its waters have not yet been diverted underground. Part of the upper basin is underlain by impermeable rocks, part is plastered with boulder clay, and part of the inner valley is lined with sediment deposited by the stream itself. In these conditions, surface flow is maintained. At an earlier time the waterfall stood farther downstream, at the very edge of the plateau. But as the stream has cut down and back, the fall has been reduced to an irregular cascade and the valley has been deepened into a slot.

It can confidently be assumed that, in the future, sinks will be opened in the bed of Gordale Beck on the upstream side of the gorge. The cascade will then run dry. At that stage it will be comparable to the disused waterfall at Malham Cove, near by. Perhaps a fitter comparison is between Gordale Scar and Cheddar Gorge, another steep-sided limestone valley cut by a vanished river.

When a stream is diverted underground by sinks in its bed, valley-deepening is controlled by the terminal sink. If a large sink remains in use for a long time, the valley upstream can be greatly deepened as the lip of the sink is worn down. In these conditions, the surface-river flows into a trench which, descending towards the sink, ends abruptly there at an enclosing wall.

Dry valleys remain as evidence of old river-courses on the surface, and by their means the former pattern of surface-drainage can be reconstructed. But there is no reason to suppose that underground drainage follows the same lines, even when the bottom of a dry valley is pitted with disused sinks. As already mentioned, underground drainage at Water End goes eastwards, whereas the outlet at the surface is to the northwest. In really strong limestones, where extensive systems of cavities occur, it is quite usual for the direction of underground flow to be controlled mainly by the attitude of the rocks (Fig. 19).

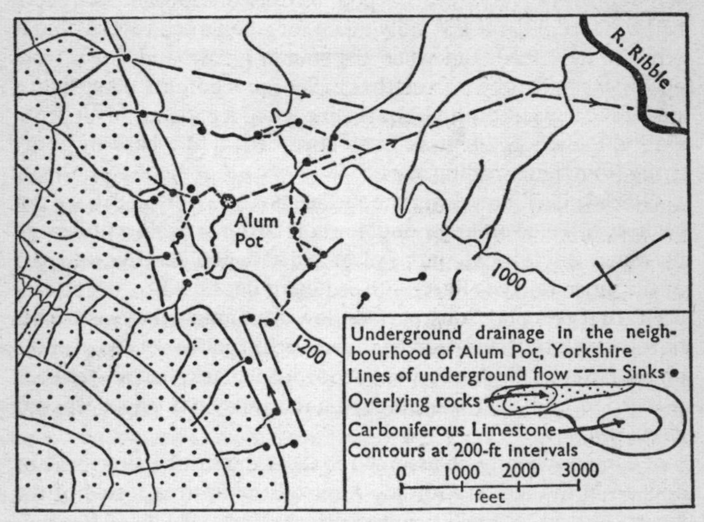


Fig. 19 Underground flow (re-drawn after Dwerryhouse)

Among the underground features of limestone country, large caverns fantastically decorated with deposited lime are the most memorable (Plate 21). Numbers of them have been turned into showplaces, where guides supply grossly misleading and unhelpful comments on their origins. In actuality the development of limestone caverns is not fully understood.

There is little difficulty in accounting for the pipes which lead steeply down from the surface through flat-lying limestones. Small underground tunnels are readily explained by solution along planes of bedding and of jointing. But enormous caverns pose an altogether different kind of problem.

Most of the origins suggested for caverns fall under one of three heads. Some writers believe that the openings have been made in the zone between the surface of the ground and the water-table – the zone in which downward percolation is dominant (Fig. 20*b*).

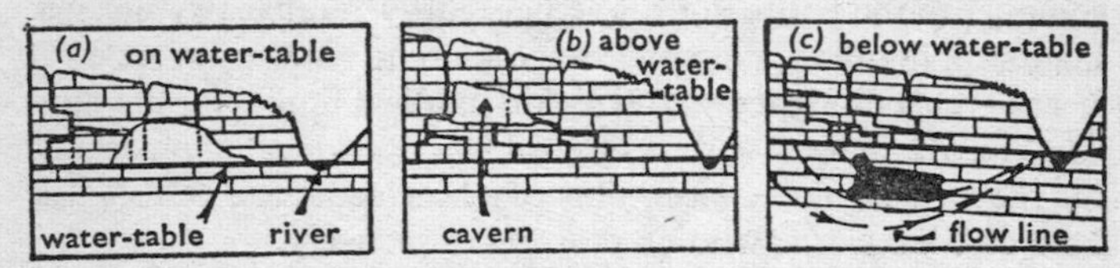


Fig. 20 Suggested origins of limestone caverns

Others contend that the initial openings are made close to the water-table, on the grounds that water is always present at the relevant level and that its flow is more vigorous there (Fig. 20*a*). The third possibility is that the limestone can be extensively dissolved below the level of saturation (Fig. 20*c*).

Percolating water can attack limestone because it is charged with carbon dioxide. The principal source of $CO_2$ is not the free atmosphere but the air contained in soil, where concentrations of the gas can be nearly a hundred times as great as in the air above ground. Underground water in limestone regions normally contains about 400 parts per million of $CO_2$, and a value of 2000 parts per million has been recorded. It appears certain that solutions of this description could act on limestone rocks well below the limit of saturation. As so often happens, however, study of actual cases reveals that developments have been complex – in particular, that no cavern system can be understood unless allowance is made for significant changes in local conditions. The Carlsbad Caverns of New Mexico are located in a region where the present climate is notably dry

One must suppose that the caverns were opened in conditions wetter than those now prevailing. Sweeting, in a study of the Pennine caverns, shows that the floor-levels of the great cavities arrange themselves in horizontal series, which accord with the heights of the erosional platforms identified on the surface. In other words, the development of caverns in the Pennines appears to have been regulated by the deepening of ordinary valleys, which in turn controlled the level of the water-table under the adjoining hills.

Limestone features reach their extreme development, on a regional scale, in the karst country of Dalmatia. There, despite rainfall rising to 200 inches a year, the land looks like a desert. In some parts the entire surface is pocked by numberless sinks of huge size; in others, nothing is to be seen but rotting pinnacles. It is a shocking thought that this region supplied timber for the ships of Venice in that city's days of greatness. Reckless felling, exposing the surface to rain-wash, allowed all surface soil to vanish. The land now lies waste, with no prospect of recovery.

*

The patterns produced by the adjustment of drainage to geological structure are widely used in photogeology. Structure and outcrops can be read from stereoscopic air photographs. Simple structures in sedimentary rocks are often obvious at a glance. Direction and angle of the dip of the rocks can be measured. Limestone terrain, with its distinctive pattern of vanishing streams, can be recognized at once. On rocks other than limestone, ground checks are necessary. Field parties must travel across country, or must go from point to point by helicopter. But the drainage-pattern – often ring-like on eroded domes, rectangular on terrain affected by cross-faulting, tree-like horizontal strata, and so on – frequently supplies an essential guide.

A totally different, and theoretical rather than immediately practical, approach to drainage nets concerns network geometry. This, in spite of what might first be thought, deals not with the drainage pattern but with the internal relationships of networks. It is wholly quantitative. Such terms as *widely spaced*, *close-spaced*, *open-textured*, and *fine-textured*, are in their application to river systems completely matters of judgement, or even matters of opinion. They mean different things to different people. Indeed, even the term *river* means different things to different people. A small boy in the dry Australian outback asked his teacher what a river is called when it has water in it.

The initial impetus to the modern treatment of network geometry, involving measurement throughout, was supplied by R. E. Horton in 1945. Horton made what he called an hydrophysical approach to quantitative morphology. That is, he showed how to analyse network geometry in a numerical fashion, and started to explore the means of relating his results to hydrologic theory. Thirty years later, the implications of his ideas were still being pursued.

The techniques involved are statistical. The fundamental principle is extremely simple. It amounts to a statement that rivers grow like trees.

In some ears, *statistics* is a dirty word. We should do well, however, to recall Weldon's generalization that the basis of science is statistical, and that statistical statements are not rules – they are merely what happens. We ought also to remember that the purpose of statistical analysis is to reduce variable behaviour, such as that of rivers, to some kind of manageable pattern.

Long before Horton wrote, statistical approaches to the definition and analysis of relief were known. Hydrologic theory was also well established. But Horton supplied the essential idea, that streams can be numbered upward according to their position in a drainage net.

The most widely used ordering system, slightly modified from Horton's, is illustrated in Fig. 21. A first-order stream is a stream of lowest rank. It is a headstream with no branches. Two first-order streams unite to form a second-order stream. Two second-order streams unite to form a third-order stream: and so on. Promotion in rank happens only when streams of equal rank unite. A third-order stream can receive any number of first-order or second-order tributaries without being promoted to the fourth order.

Horton listed fifty-seven variables as deserving study. Of these, in addition to stream order, the basics are stream number, stream length, stream slope, and the bifurcation ratio. When values for these variables are determined, an orderliness appears which is quite independent of drainage-pattern.

The fictional network in Fig. 21 contains 14 first-order streams: 14 is the stream number for the first order. The second order contains 6 streams, the third order 2 streams, and the fourth order one stream. It is obvious that number must decrease as order increases.

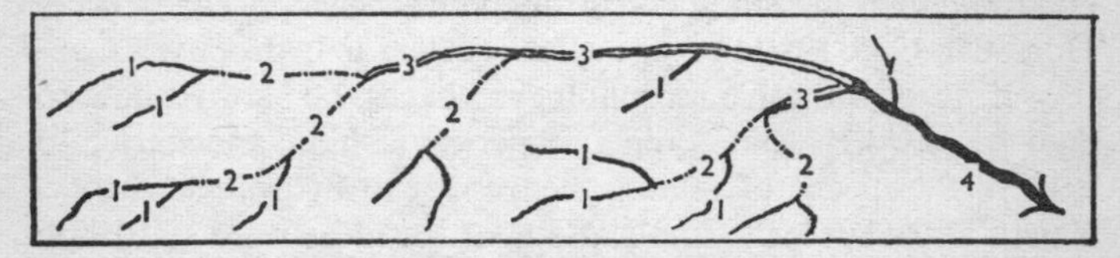

Fig. 21 Stream order

When stream number is plotted against order, for a given basin or for a region, the points tend to arrange themselves in straight lines (Fig. 22a). In the diagram, the order scale (left) is arithmetic, the number scale (bottom) is logarithmic. The appearance of a nearly straight line through the plotted points means that, as order decreases, the number of streams multiplies. The multiplying factor is the *bifurcation ratio*, the index of branching. A distinct difference appears between the Ozark Plateau, with 198 first-order streams, 38 second-order streams, and 10 third-order streams, and the Gulf Coastal Plain, where the corresponding numbers are 96, 27, and 8.

Similarly with stream length. This is the average length of stream per order. In the Ozark Plateau, mean length is 0·10 mile for first order, 0·132 mile for the second order, and 0·368 mile for the third order. Corresponding values for the Gulf Coastal Plain are 0·260 mile, 0·427 mile, and 0·844 mile. As with number, a numerical comparison can be made between the two regions. But the really important matter is that mean stream length tends to multiply by a constant factor as order increases by addition. We are dealing with a mode of growth represented also by the accumulation of compound interest and by the population explosion.

Similarly, again, with stream slope. Slope tends to decrease by division as order increases by addition, just as stream number does. Here is an explanation – or possibly only a description – of the fact, explored further in a later chapter, that slope tends to decrease in an orderly fashion in the downstream direction. The interrelationships of order, number, length, and slope can all be expressed for a given network by equations. To some, equations are even more offensive than columns of data, or than graphs drawn from the data. However, it remains true that equations are merely symbolic ways of stating the forms, positions, and slopes of lines drawn on graphs.

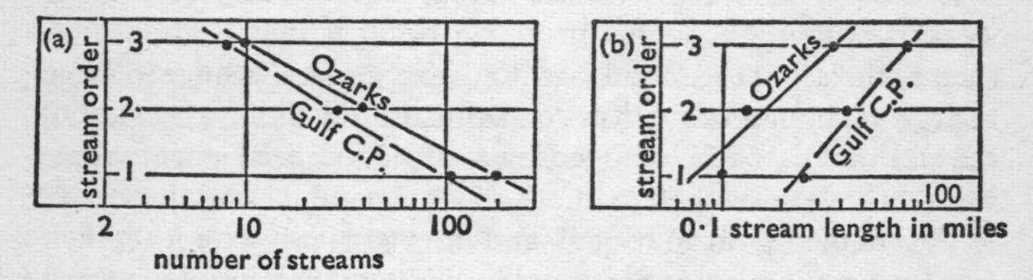


**Fig. 22 Order, number and length of streams (adapted from Strahler)**

Looking back at the qualitative, subjective, ways of describing drainage texture, we can now see that the analysis of network geometry provides us with the means of quantitative comparison. When the total length of channel in a given basin is divided by the drainage area, the result is the drainage density. This, by definition, is the reciprocal of the mean distance between channels. The average length of overland flow – the distance that water running across the surface travels before a channel forms – is half the reciprocal of the drainage density, that is, half the average distance between one channel and its next nearest neighbour. The value of drainage density can be useful in assessing liability to soil erosion.

Two points remain outstanding: the reason why streams branch at all, and the pursuit of the theory of branching systems. Branching is in fact readily understandable. A stream, growing headward, will throw out a branch as soon as there is enough extra territory to provide enough runoff to cut a channel. The branch will itself branch, as soon as it invades enough territory to supply runoff sufficient for two channels. The result is the formation of branching nets, which in terms of order, number, and length are strictly comparable to what happens with trees. Readers are invited to try trees out. Plum trees are especially easy to deal with. The parallel is known to extend to the repeated subdivision of air-passages in the human lungs, which behave just like streams. It also extends to the arterial system in the human body, where artery diameter is halved at each subdivision. It seems that when Nature (whatever Nature may be) hits on a good idea, then that idea is applied in a multiplicity of contexts.

Investigation of the facts and theory of branching stream networks still continues. One probable result is that relationships, such as those of peak flood flow to drainage area, change in some regions at about the fourth order. Fourth-order basins are of about the size that could be deluged by a single localized storm. Some workers have developed methods of stream numbering alternative to the adapted Horton model, and are vigorously exploring their implications. As is now common, investigation depends increasingly on the use of computers, which are fed with observational data or are instructed to say what would happen in a given case.

One outcome, initially unforeseen, is that branching networks, geometrically similar to natural networks, can be produced by random variation of the branching system. Dice, cards, or computers will generate nets which comply with the laws of order, number, and length. Many researchers find this outcome reassuring. As with statistics in general, random variation is simply what happens.

We infer then that behaviour – of stream nets, of particle size on beaches and on stream-beds, of angles of hillside slope – is being correctly defined. Normal and log-normal distributions of frequency enable us to answer the question *What happens?* We can then go on immediately to address the question *Why?*, involving ourselves in the study of process.

CHAPTER FIVE

# *Volcanoes and Earthquakes*

EARTHQUAKES and volcanic eruptions make news. On occasion, they also make films. The San Francisco earthquake of 1906 has supplied the high note of drama in more than one Hollywood production. During the 1970s appeared the first movie to portray a fictional future shock in the California coastland. Complete with vibratory sound, this production cracked susceptible buildings. Scientists now keep careful cinematic records of new volcanoes, such as Paricutin in Mexico and Surtsey off the Iceland coast.

On the human time-scale, earthquakes and volcanoes are so catastrophic that individual disasters assume a legendary character. The Shensi (China) shock of 1556 is reputed to have killed more than 800,000 people. The eruption of Vesuvius in A.D. 79, the first to be recorded in detail, buried the cities of Herculaneum and Pompeii. The detonation of Krakatoa in the East Indies, in 1883, was heard in Africa. Ejected dust, travelling through the higher air, produced brilliant sunsets as far away as Europe, and circled the globe three times before it finally settled.

When the Japanese volcano Bandai-san exploded in 1883, 3 million tons of rock were blown to dust. The Assam earthquake of 1897 caused serious damage in an area of 150,000 square miles. At Messina in 1908, 100,000 people were killed in a few minutes. At least as many died in the Japanese shock which devastated Tokyo in 1923. As many again were injured, and half as many were reported missing. Parts of coastal Alaska suffered severe earthquake damage in 1964. Also during the 1960s, a major shock in Chile caused the earth to ring like a bell.

Small wonder, then, that sudden disasters, whether seismic or volcanic, should be thought supernatural. They are noisy and destructive. Until recent years they have always been unpredictable. They seem to exemplify the wrath of the gods. Indeed, the Lisbon earthquake of the eighteenth century, the very Age of Reason, was widely accepted as heralding the end of the world.

Mere humans can be excused for being repelled by noise, devastation, and apparent capriciousness. At the same time, many volcanoes possess exotic names – Cotopaxi, Orizaba, Stromboli, Papandayang, Hekla, Etna, Kilimanjaro, Hualalia, Mauna Loa, Fujiyama. This last, indescribably beautiful and pictorially the best known of all, sets the type for volcanic cones. It is understandably sacred.

In the ancient myths of the Mediterranean, Vulcan's smithy was located in the Lipari Islands – to be precise, in the mountain Vulcano which supplies the type-name. The lake-filled crater of Avernus, not far from the modern Naples, figured as the entry to the underworld. It is a reasonable guess that the Avernus myth related originally to a deep hole which emitted volcanic gases. Here, surely, was the bottomless pit, and the type locality of hell-fire and brimstone.

A fiery hell somewhere beneath the surface of the earth represents an attempt to explain observed facts in terms of early religion. But there is little space for flame-filled chambers in the present geological scheme, which requires the rocks deep underground to be tightly packed. At the same time, it is extremely probable that rock-temperatures at depth are very high indeed. Scarcely anything is known of this matter by direct observation, but inferences may be drawn by indirect means. The question is important in the present context, for volcanoes depend on a supply of heat to keep them going.

Measurements in deep boreholes, in mines, and in tunnels through mountains suggest that rock-temperature increases with depth at about 1° C for every 100 feet of descent. In practice there are rather wide departures from this rough figure, but as an average value it seems to be of the right order. If such an increase were maintained right down to the centre of the earth, the temperature there would be about 200,000° C – a figure high enough to alarm any astronomer or geophysicist. But the increase is not maintained. At the shallow depth of 60 miles, it is thought, the rate of increase has already fallen to 1/60 of its value near the surface. Even so, rock-temperatures 60 miles down probably stand at 1500° C – quite high enough to melt any rock *at atmospheric pressure*. Here is the reserve of heat by which volcanoes are kept in action. It remains to be

1. Layers in soil on Chalk, North Downs

2. Weathered stone effigies, Oxford

3. Rock-face and scree, Western Ireland

4. Weathering-pinnacles, New Zealand

5. Weathering-pinnacles, Utah

6. Rotting limestone, North Pennines

7. Rotting dolerite, Fife

8. Sandstone rotted by salt-spray, Donegal

9. Section through soil to rock, Northants

10. Slumping, Dorset coast

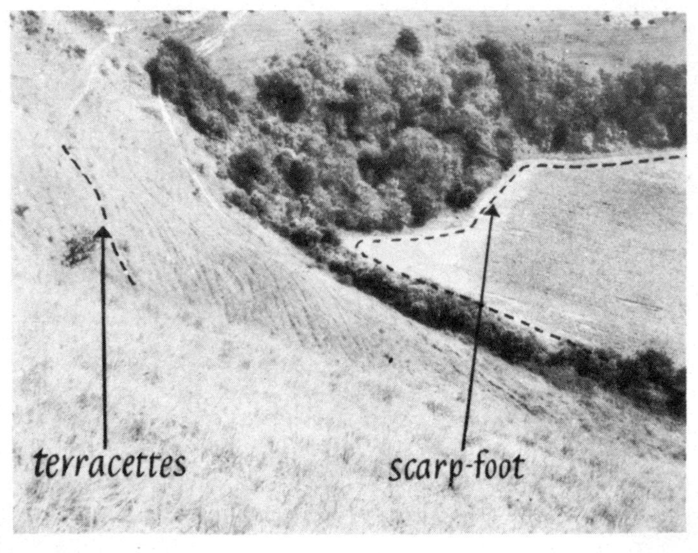


11. Terracettes on Chalk, North Downs

12. Dry valley in Chalk, Sussex

13. Gullies, California

14. Dry valley in Chalk scarp, Chilterns

15. Old sinks in Chalk quarry, Herts

16. Dry valley in Cotswold Limestone

17. Dry valley in Carboniferous Limestone, North Pennines

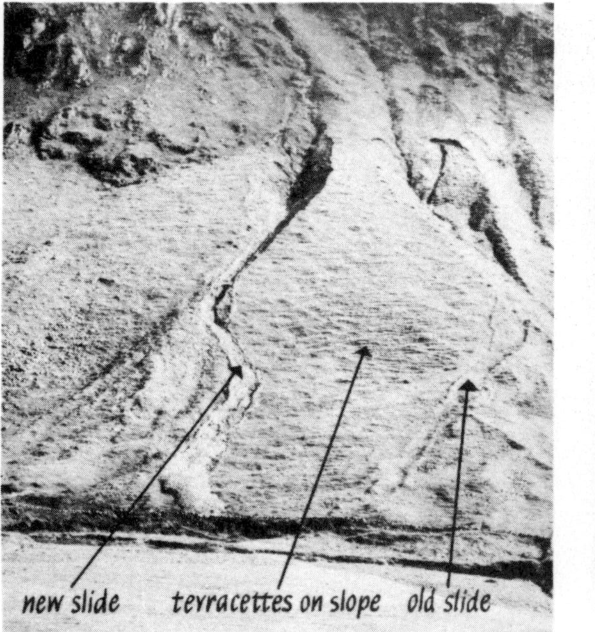


18. Mudflows, Glengesh, Donegal

19. Sink in Carboniferous Limestone, North Pennines

20. Gordale Scar – gorge in Carboniferous Limestone

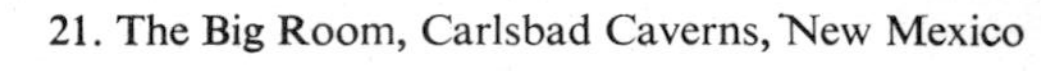
21. The Big Room, Carlsbad Caverns, New Mexico

22. Columns in basalt, Staffa, Argyll

23. Paracutín erupting

24. Dissected basalt country, Hawaiian Islands

25. Broken explosion-crater, White Is., Bay of Plenty, N.Z.

26. Vertical air view of African cross-faults

27. Pothole in stream-bed

28. Ebor Falls, N.S.W., Australia

29. Meandering valley of the Evenlode, Cotswolds

seen how heat and molten rock are brought to the surface, and why the vents through which they escape are localized and small.

A word, first, about the origin of the heat. Part of it represents the primitive heat of the earth, part is generated by radioactivity. At a depth of 60 miles the two parts are thought to be roughly equal, but at a depth of 6 miles radioactivity contributes three-quarters of the whole (Fig. 23). Were it not for radioactive heat, volcanic outbreaks would certainly be far less numerous and far less frequent than they are, and might even fail to occur at all. But as matters

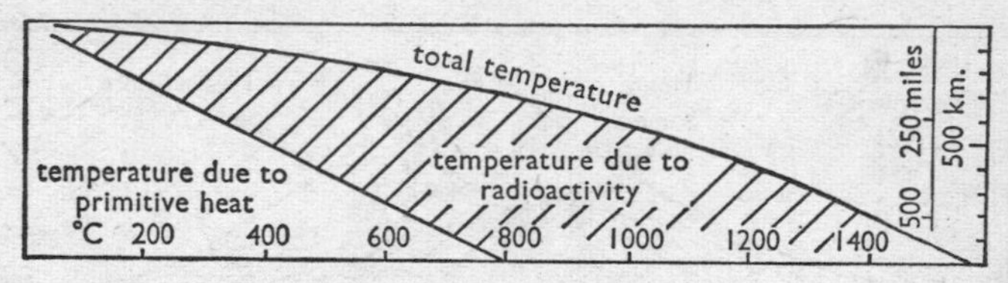


Fig. 23 Earth-temperatures

stand, heat, gases, and molten rock can and obviously do escape.

Volcanic activity is intermittent as well as localized. Some volcanoes burst into violent life after long dormant spells, and all are doomed eventually to become extinct. The relevant characteristics are best explained by regarding volcanoes as located at the ends of cracks in the crust – cracks deep enough to traverse the crust from top to bottom. Why cracks should form in the first place is a conjectural problem, but assuming a crack to exist in a crust some 40 miles thick, there is no difficulty in understanding the ascent of molten rock. The local difference of pressure at the bottom of the crack is 17,000 times as great as the pressure of the atmosphere at sea-level – quite enough to force molten rock into the crack, to open a fissure right through the crust, and to raise the molten rock to the surface of the ground.

Normally, the rock immediately beneath the crust is held rigid by pressure. When pressure is reduced, as, for instance, by the first opening of a crack in the firm crust above, the rock can liquefy. In the molten state it is known as *magma*. During ascent, magma becomes lighter through expansion and is made mobile by the release of liquids and gases. In extreme cases it can be distended into a froth. For these several reasons, it is capable of pouring out at the surface as lava. But while outflows of lava can thus be explained, volcanic explosions and the building of volcanic cones have still to be accounted for.

An important problem is the typically small size of central vents. Hundreds of old vents are known to geologists, in the guise of pipes plugged with frozen lava or with masses of broken fragments. Their average diameter is well under 1000 feet. Through vents of such small dimensions come the dust, rubble, and lava which accumulate in cones thousands of feet high and miles across at the base.

Fig. 24 Vesuvius and Monte Somma

In May 1945, when it was at the height of its activity, Paricutin was estimated to discharge 100,000 tons of lava a day, in addition to quantities of debris. In the 2 years of its life, this volcano had already made a cone 1500 feet high, 10 miles across, and 150,000 million tons in weight. All the material had come through a single narrow throat.

It is easy to realize that central vents can be plugged if the lava cools enough to freeze. But if heat continues to rise from below, energy is built up below the frozen plug. On occasion, a plug is forcibly displaced and eruption violently renewed. Long spells of dormancy, alternating with vigorous outbursts, are due to intermittent plugging of a volcano which is not yet exhausted. Vesuvius exploded in A.D. 79 after a lengthy inactive period, blowing off the top of its cone and leaving a huge explosion-crater rimmed by the sharp ridge of Monte Somma (Fig. 24). In 1902 a cooling plug was actually extruded from the vent of Mont Pelée, on the French West Indian island of Martinique. An eruption was closely followed by an unusually horrifying event: Mont Pelée discharged an incandescent cloud of volcanic dust. Rushing down the side of the cone at 200 miles an hour, the cloud laid waste the town of St Pierre, boiled the water in the harbour, and sank the anchored shipping. Subsequently, the slowly freezing plug appeared as a rising column at the top of the mountain, but soon began to crumble as it was attacked by weathering-agents.

By contrast with Vesuvius, Mont Pelée, Krakatoa, and Bandai-san, some volcanoes are remarkably quiet and also remarkably regular in their activity. Stromboli rarely blows anything beyond the lip of its crater, Solfatara erupts nothing but gases, and the lava-lakes of Hawaii simply rise and overflow their margins. Contrasts in behaviour seem to be due partly to differences in the supply of heat and partly to differences in the constitution of the lava. In addition, there are probably differences in the power of certain processes which operate in the pipes.

Stromboli and Solfatara are thought to be near the end of their volcanic lives. Their reserves of heat seem to be running low. But even so, there is enough heat available to cause explosions if the vents were not kept open. Prolonged activity unbroken by plugging and explosions depends on the maintenance of a free outlet to the surface. It seems highly probable that vents can be kept open, or reopened after plugging, by a kind of blow-piping – that is, by the action of hot gases coming off the magma. In favourable circumstances, the tendency for a plug to freeze can be offset by fluxing. In other circumstance, fluxing weakens a plug which has sealed the vent, and activity is abruptly renewed.

Of the erupted gases, steam is by far the most plentiful. It amounts to 70 per cent of the gases discharged by the Hawaiian vent of Kilauea, with carbon dioxide and sulphur dioxide accounting for most of the remainder. Paricutin released 16,000 tons of steam a day at its most active (Plate 23). In all probability, steam supplies the enormous pressures indicated by volcanic explosions – explosions far greater, in some cases, than any man-made detonation yet recorded.

Volcanic steam is thought to be released by crystallization. The way in which the process operates may be illustrated by reference to the mineral albite ($Na_2O.Al_2O_3.6SiO_2$), a common constituent of solidified crystalline rocks. At a temperature of 1100° C, and under the pressure appropriate to a depth of 7500 feet below the surface, albite holds about 4 per cent of water. Crystallization begins when the temperature falls to 960° C. When it is down to 820° C, half the albite has already crystallized. The remaining liquid contains nearly 10 per cent of water. Already, because of excess water, the pressure has risen to 3000 times atmospheric pressure, and will be double again if crystallization continues.

Deep-seated crystalline rocks have frozen at temperatures of 300° to 600° C. Lesser bodies of intrusive rock froze at 500° to 700° C. Lavas freeze at 700° C or above, but much higher temperatures are reached on the surfaces of lava-flows. Measurements on flowing lava in Hawaii gave values of 1000° to 1200° C on the surface, as opposed to some 800° C at depths of a very few feet. The extra heat indicated by high surface-temperatures was released by burning gases. It helps to explain why certain lavas – including those of the Hawaiian volcanoes – can flow long distances before freezing to a standstill.

The least mobile lavas are of the acid kind – lavas which, in the solid state, contain much free quartz ($SiO_2$). Acid lavas have a high viscosity, freeze at high temperatures, and do not flow far. They are apt to plug the vents in which they rise, and are typical of explosive, cone-building volcanoes. Basic lavas – lavas deficient in quartz – are mobile. They run in long tongues or in wide sheets, at speeds measurable in miles an hour, from big central vents such as

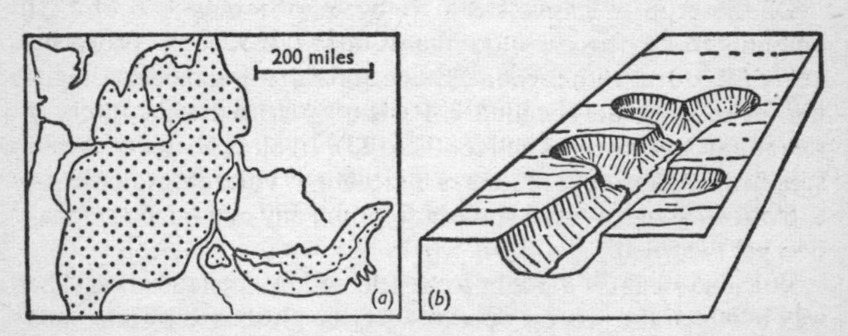


Fig. 25 (*a*) Lavas, in part basic, in the northwest U.S.A.
(*b*) Valley-ends in basalt

those in Hawaii or from long fissures such as the Laki fissure in Iceland. In 1783 lava poured from the Laki fissure over a length of 20 miles, flowing outwards to a distance of 30 miles on one side and 40 miles on the other.

Here, in the varying mobility of lavas and in varying modes of eruption, is the basis of a division of volcanic land-forms into the three classes of volcanic cones, lava-domes, and basalt sheets. These three classes will now be discussed, in reverse order.

In the basin of the Columbia river, in the northwestern U.S.A., repeated fissuring allowed 60,000 cubic miles of lava to flow widely over a rugged landscape. At least 100,000 square miles, and possibly as much as 200,000 square miles, were inundated by molten rock (Fig. 25*a*). On cooling, the quartz-poor lavas froze into the dark, fine-grained rock called basalt. In comparison with the characteristically huge extent of plateau basalts – other great spreads are known in Basutoland, in Peninsular India, in South America, and in northeast Ireland – the thickness of individual flows and the width of individual fissures are minute. An average sheet is about 40 feet thick, an average fissure about 5 feet wide. But just as repeated outpourings can create multiple sheets with total thicknesses of thousands of feet, so can repeated reopening of fissures create very thick dykes.

Where they are little dissected, basalt sheets display wide, subdued, and monotonous surfaces. Even where rivers have established themselves, valleys are widely spaced, for plateau basalt is highly permeable. Not only is the rock itself well jointed – its surface is usually coarse and clinkery, totally unable to retain surface-water. As in some limestone country, powerful springs emerge from valley-walls. The strongest are fed through tunnels in the basalt, where still-molten lava drained away beneath an already-frozen layer above. The Rogue river, in Oregon, vanishes down a basalt tunnel, reappearing only when the tube is cut across by the surface of the ground. At Burney Falls, California, a basalt spring discharges 100 million gallons a day.

Main valleys in the basalt country of the Columbia basin are deep and sheer-sided, both because the rivers are cutting vigorously downwards and because the well-marked vertical jointing of the rock lends itself to the development of sheer walls. Minor valleys, with their steep walls slowly retreating before the attack of weathering, run up to boldly rounded heads (Fig. 25*b*). Sections at the edges of basalt sheets frequently reveal columnar jointing (Plate 22). Each column is bounded by tension-cracks which formed in the cooling, solidifying, and contracting lava. Where – as commonly happens – the sheets lie roughly horizontal and the columns are vertical, the effect is one of barbaric masonry.

Lava-domes, the second class of volcanic land-forms, are best known from the Hawaiian Islands, where a whole chain has been formed on the line of a 1600-mile fissure in the ocean-floor. Five volcanoes compose the island of Hawaii itself. Mauna Kea rises 13,784 feet above sea-level, and Mauna Loa 13,679 feet. Since the foundations of the Hawaiian volcanoes rest on the deep ocean-floor, the highest summits belong to volcanoes more than 30,000 feet high.

Stages in the growth of the Hawaiian volcanoes have been summarized by Stearns, who considers that the fissure opened 25 million years ago. Most of the subsequent time was needed to bring the original submarine volcanoes up to sea-level. On their first appearance above the surface, the cones were formed mainly of weak, easily-eroded ash, but this was soon veneered by lava and growth was able to continue.

The main bulk of the domes consists of basalt, which in the molten state was highly fluid. It was extruded principally from fissures. Being very porous, and flowing out at short intervals, this early basalt prevented stream-erosion and valley-cutting from taking place. Mauna Loa and Kilauea represent the next stage of development, in which a volcano gradually collapses near the vents to form depressions of various kinds, including the shallow, rounded basin called a *caldera*. The caldera of Kilauea contains the famous lava-lake. Lava-flows in this stage still consist of very fluid material, which forms ropy patterns on its freezing surface. Lava can accumulate thickly in the sunken hollows, but if it is ponded back it obviously cannot flow down the sides of the volcano; in these sides, therefore, valleys are eroded. In time, the hollows are completely filled with lava, and the volcano grows upwards again. Its sides are steeper than at earlier times, for the lavas now extruded are thick, slaggy, and rather viscous; in addition, much fragmentary material is blown from the vent. Mauna Kea is at present in this stage of development.

Later still, with volcanic activity past – or, at the most, feeble – hundreds of radial valleys develop on the sides of the domes. The terrain of highly dissected slopes can only be called savage (Plate 24). Downcutting is more effective because stream-gradients are continually steepened as the cliffs recede. On shores exposed to the trade winds – those facing the northeast – cliffs rise as high as 3000 feet.

Fig. 26 Radial streams on volcanoes: (*a*) Mt Egmont. Contours at 1000 ft intervals: land over 4000 ft stippled

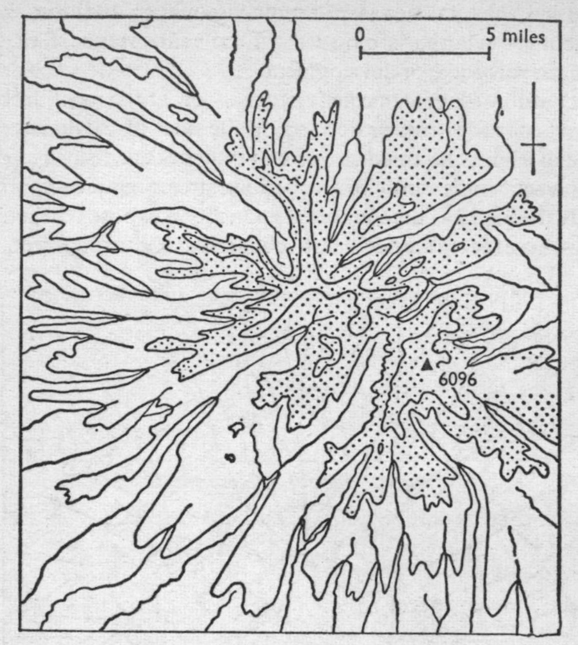


**Fig. 26 Radial streams on volcanoes: (*b*) Cantal. Contours at 1000 ft intervals: land over 4000 ft stippled**

Two main differences separate basalt sheets from lava-domes of the Hawaiian type. The sheets ran broadly out from long fissures, and these fissures presumably tapped a single reservoir of molten rock. The most fluid of the later Hawaiian lavas had an initial slope, and poured from a fairly well-localized vent; each of the main Hawaiian vents is – or was – fed from a separate reservoir.

Volcanic cones result from deposition at still steeper angles, from the eruption of more viscous lava, and from the activity of vents which are strictly localized. Typically, they are drained by radial streams (Fig. 26). If nothing radical happens to disturb the radial pattern it may persist for a very long time indeed. For example, the extinct Cantal volcano in France has a very lengthy history both of activity and of erosion. First erupting about 15 million years ago, it was still emitting enough lava and dust 7 million years later, to smother 1200 square miles of country. Although now long extinct, and deeply dissected, it retains a distinctly radial pattern of drainage (Fig. 26*b*).

In considering the detailed forms of single volcanoes, much allowance has often to be made for the local effects of lava-streams. Valleys furrowing the sides of a cone guide the course of lava which bursts through the crater-wall. Thus valleys can be plugged, and stream-courses directed along the lines of former watersheds (Fig. 27).

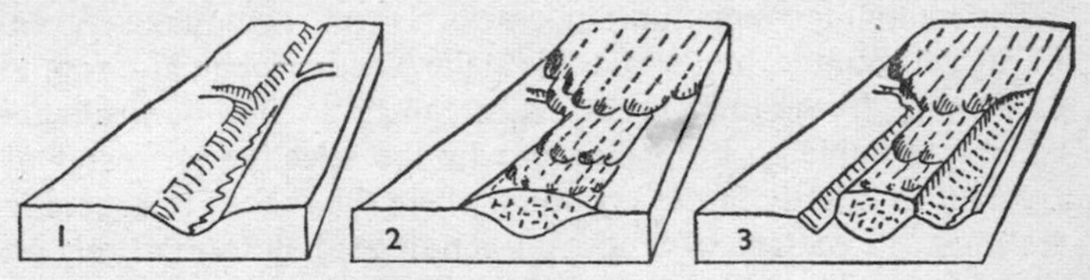


Fig. 27 Plugging of valley by lava

Given time, stream-erosion will attack the centre of a cone, and can gut it completely. A large hollow at the centre of an old volcano (Plate 25) can, therefore, originate in more than one way. It may be due simply to selective erosion, to explosion – as with

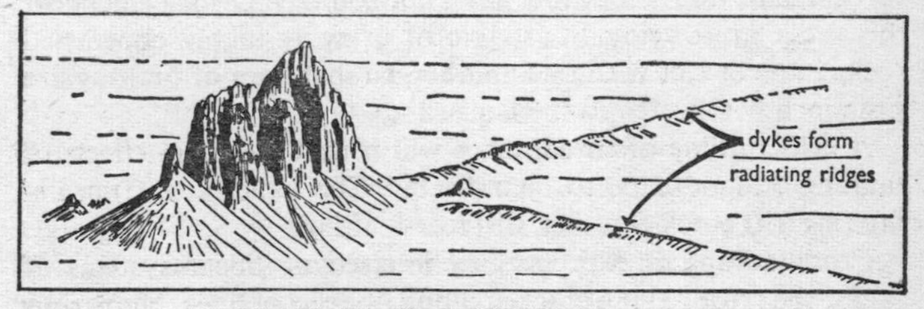


Fig. 28 Ship Rock, New Mexico

Vesuvius – or to explosion followed by subsidence, as with Krakatoa. But if erosion alone is at work, the lava-plug in the lifeless vent may prove so resistant that it rises in the form of a pinnacle. The well-known Ship Rock of New Mexico, which is just such a pinnacle, towers at the focus of ridges based on radiating dykes (Fig. 28).

*

Earthquakes are coupled in this chapter with volcanoes both on account of their disastrous character and because they too are associated with weakness in the earth's crust. But the tremors which accompany volcanic outbreaks are negligible by comparison with the devastating shakes of great earthquake-shocks. Some of the implications of earthquakes will be noticed later. For the present, attention will be confined to land-forms of regions liable – or formerly liable – to frequent earthquakes: in other words, to the features developed on lines of faulting, where the crust of the earth is cracked through.

Differential movement along cracks sets off earthquake-waves. In rugged country, huge landslides take place as incidental results of shaking. In the longer run, faults constitute lines of weakness which are liable to be picked out by the selective processes of erosion. In heavily faulted country where the rocks are generally resistant, the pattern of landscape can depend almost entirely on the pattern of faulting. Again, although single movements rarely exceed a few feet, repeated movements can effect displacements measured in miles. The greatest known movement at one time, recorded from Alaska, amounted to 47 feet at the maximum; against this should be set aggregate movements of 2 miles in the vertical sense and 50 or more miles horizontally. Displacements of this order are obviously capable of bringing highly contrasted rocks into contact with one another, so that lines of faulting are prominently reflected in the landscapes.

A brief outline of three topics will permit the main effects of faulting on landscape to be indicated. The topics are: (i) cross-faulting, (ii) block-faulting, (iii) tear-faulting.

Cross-faulting is expressed by intersecting linear systems of faults, the crust having cracked along two sets of lines. Numerous examples are available from parts of Africa, where straight valleys have commonly been eroded along the fault-lines (Plate 26). Between the trenches of the fault-guided valleys the landscape is extremely subdued, having been levelled in some previous cycle of erosion or reduced to broad pediments in the present cycle. The essential feature of the valley-pattern is an austere angularity – an outcome of close adjustment between drainage and structure. Despite severe glaciation – or perhaps because of it – the valley-pattern of the Scottish Highlands similarly corresponds closely to a network of intersecting faults.

Block-faulting involves tilting of the fault-bounded blocks. Descriptions of block-faulted areas relate to terrain where the landscape-pattern is influenced by tilting, either directly or indirectly. Some single blocks are large enough to be classed as regions in their own right – the Sierra Nevada of California, for instance, extends 20,000 square miles. Tilted blocks of varying dimensions underlie 125,000 square miles of country in the Great Basin of Utah, which is also large enough to require regional treatment. But it is proposed, for the sake of brevity, to concentrate on those land-forms which are eroded in the steep upturned face of a fault-block.

Despite the fact that the total displacement of a major fault represents the sum of many small movements, it is quite possible for the rate of tilting to exceed the rate of valley-cutting. Thus the uplifted fault-face remains partly, or even largely, intact at the time when movement ceases. The best illustrations of the relevant land-forms come, naturally enough, from places where recent and rapid faulting has taken place, and where the cutting of valleys is retarded under a dry climate. Dry regions have the additional advantage that their stark land-forms are not heavily masked by soil and vegetation.

Provided that its exposure is due directly to faulting, the steep edge of a tilted block constitutes a *fault-scarp*. Gullies form upon it as streams go to work; debris discharged at the gully-mouths accumulates in fans of alluvium. So long as something remains of the original face of the scarp, triangular facets are to be seen aligned on the fault (Fig. 29). Many thousands of years may be

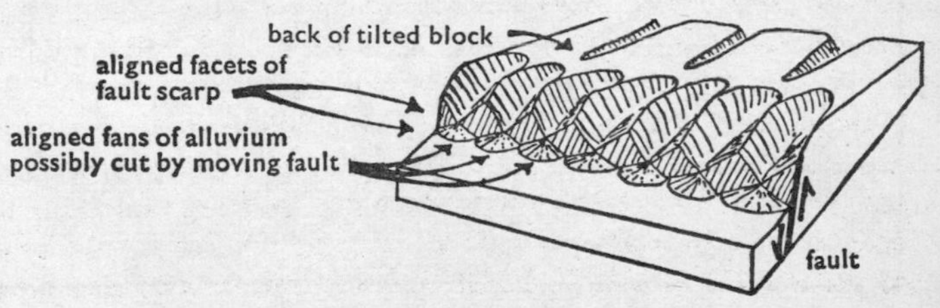


Fig 29 Fault-scarp

needed for these facets to be entirely consumed by the widening gullies, so that they bear long-enduring witness to their origins.

Developments subsequent to the entire destruction of the original face are in dispute. They also raise linguistic problems. If the steep slope can undergo general retreat, without being progressively reduced in gradient (p. 70), it is obvious that the scarp can recede from the fault-line without losing its identity. Some scarps appear to have done precisely this. They can hardly be called fault-scarps, since they are not immediately due to faulting. Nor can they be called fault-line scarps (see below), since they are not located on the fault-lines. The debate over names is at present in a state of high confusion, which could, however, be simply resolved if the steep erosional slopes in question were merely given the name of scarps, without any qualification.

Fault-line scarps are steep slopes developed on lines of faulting, not by differential movement but by erosion. Two fine examples are supplied by the southwest face of the Quantocks and by the west face of the north Pennines. In each case, faulting occurred some 200 million years ago. The depressions below the original fault-scarps were deeply filled by desert sediments. Present-day rivers, clearing away the sedimentary fills, are once more revealing the abrupt faces of the tilted blocks (Fig. 13*b*).

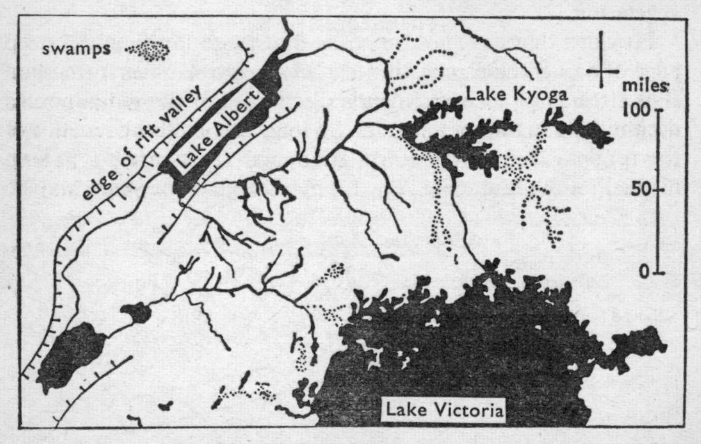


Fig. 30 African valleys flooding by tilting

A depression caused by subsidence between parallel faults is a *rift valley* (Plate 36). The greatest known systems of rifts is that of East Africa, where long, narrow lakes occupy down-faulted basins and volcanic activity accompanies fault-movement. Alongside the rifts – in some places at least – the adjacent blocks have been gently tilted, with peculiar effects on surface-drainage. Lake Kyoga, Uganda, is a large, shallow water-body with an extraordinary, frond-like plan – a result of eastward tilting, which has drowned much of a former river-valley and has reversed the flow of water in the old outlet. Lake Victoria is the outcome of more extensive flooding in a wider basin (Fig. 30).

Faults along which the movement is mainly horizontal are tear-faults. Tear-faulting is usual in the Californian coastland, especially on the line of the San Andreas Fault (Fig. 31). But it has

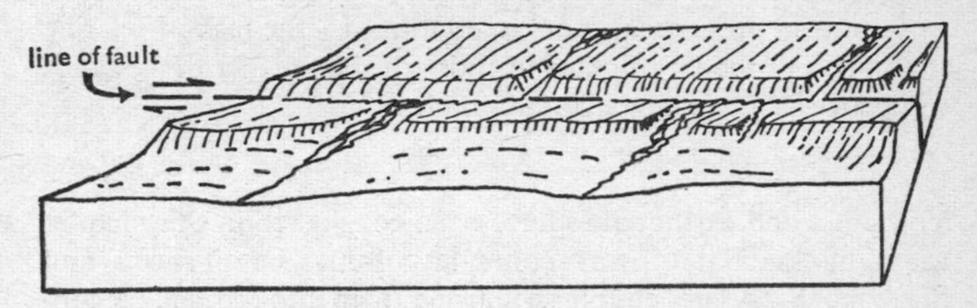


**Fig. 31 Streams offset by tear-fault**

probably been best studied in New Zealand, where tear-faults run across the grain of relief. Truncated ends of ridges have been moved, shutter-like, across the truncated ends of valleys on the opposite sides of faults (Fig. 32). Streams adjust themselves as best they can, turning into and away from the fault-line at sharp angles.

In New Zealand, where the faults are still active and where they

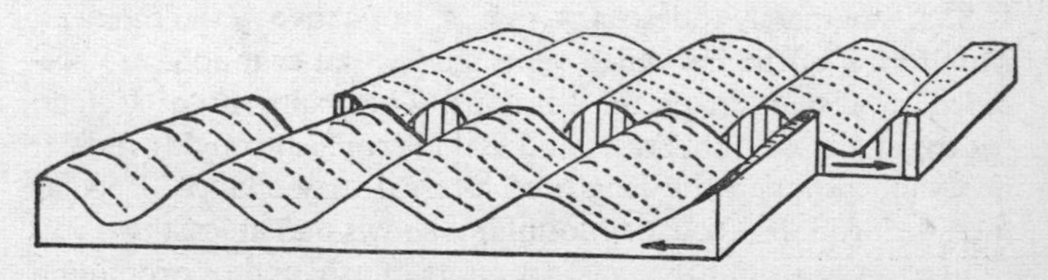

**Fig. 32 Tear-fault in dissected country, New Zealand**

are disturbing the existing streams, it is not a difficult matter to interpret the pattern of the landscape. A far severer problem is set by the Great Glen Fault of Scotland, which is considered to have shifted the northern Highlands against the Grampians for a dis-

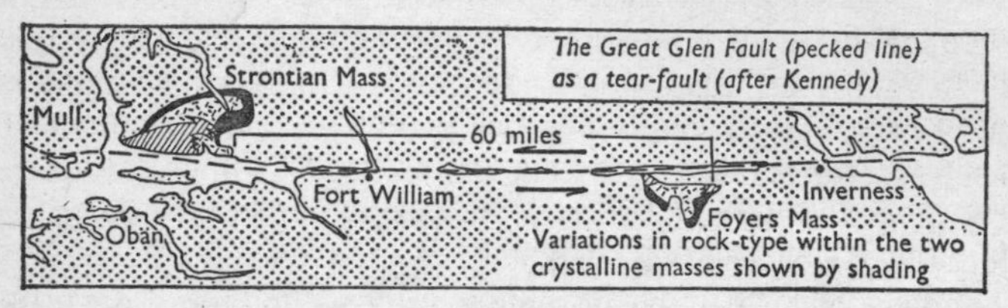


Fig. 33 The Great Glen

tance of 60 miles (Fig. 33). If this interpretation is correct – and work now in progress appears to confirm it – the task of relating the valleys of the Grampians to those of the northern Highlands will prove as interesting as it is complex.

*

Volcanoes and earthquakes have been considered in this chapter as producers of land-forms – cones, lava domes, basalt sheets, and linear cracks. A later chapter will bring them into the general context of crustal movement. For the present, attention will be directed to the prospects of predicting future eruptions and future shocks.

Although eruptions and earthquakes are localized in space and apparently intermittent and irregular in time, prediction is by no means impossible. No central volcano which has not been stripped down into ruins can be regarded as really safe. Active volcanoes pose no problems. Vesuvius before A.D. 79 was dormant. Some volcanoes, including a number in the Cascade Ranges of the U.S.A., are merely quiescent – that is, not active at the moment. Monitoring of the Hawaiian vents shows that eruptions are preceded by distension of the summits. The main rift of Iceland widens at a measurable rate; and Hekla erupts at roughly regular intervals, although the length of interval may change. Regular monitoring in the Cascade Mountains shows that at least two volcanoes, given to emitting vast amounts of ash, and of promoting ash-flows, endanger whole cities. But earthquakes remain the major hazard. One can flee from an erupting volcano. Until recently, earthquakes have struck without warning.

The infamous San Andreas Fault of California, mainly inactive in its northern part since 1906, with its sides locked together, is under increasingly intensive study. The signs of impending movement range from strain across the fault to differential tilting of quadrants defined by the line of the fault and an arbitrary perpendicular. Now comes a philosophical, moral, and ethical problem: should potentially endangered people be warned, at the risk of panic? The current situation is that the 1906 shock involved lateral displacements of crustal blocks by as much as 20 feet. The pressures built up since 1906 are enough to cause another displacement of at least 10 feet. A reasonable forecast is that the next major shock, bringing horrifying destruction, will occur by about the year 2000 – possibly sooner, possibly tomorrow, certainly before the year 2150.

We know that the injection of fluids into areas prone to active faulting can cause crustal tensions to be released. However, matters have gone too far in the San Francisco area for the release of tension to involve anything less than a fearsome earthquake. Hence, of course, the repeatedly revived myth that part of California will slide into the sea. It will not. But it will violently shake.

CHAPTER SIX

# *Wearing Down or Wearing Back?*

MUCH of what has been said so far relates to selected examples of actual landscapes, each example illustrating some particular aspect of earth-sculpture. The present chapter is intended to be general rather than specific. Its theme is the systematic course of landscape evolution, between the time when uplift exposes rocks to weathering and erosion, and the time when almost all the uplifted land is denuded away.

No apology is required for the introduction of theoretical or conceptual matter. Just as there are limits to what can be done in the field, by means of random and casual observation, so there are limits to what can be achieved by means of description, when this is unrelated to general principles. This statement holds, no matter how careful and detailed the description. But since there will in fact be occasion to refer to field examples, readers who may be averse to theoretical argument or to broad conceptual sketches are urged to press on and see what happens.

If the study of earth-sculpture is to be truly scientific, it must have a firm basis of valid theory. This basis will be supplied by statistics, applied mathematics, chemistry, and physics – and also, as we come increasingly to understand the impact of organisms on the earth's surface, by biology. When theory is in dispute, as it often can be in natural science, then original field observations need to be rechecked, and fresh evidence needs to be sought. Very little of the earth's landscape has yet been examined carefully and in detail. Inference about the development of landscape depends strongly on the character of terrain for which general conclusions are derived. Inevitably, researchers tend to regard as standard those landscapes with which they are most familiar.

Also, all but a few pioneers are powerfully affected by the fashion of their time. Only a tiny minority attempt to provide conceptual frames into which observations in general can be set. Of this minority, only a fraction succeed. Thus it came about that the so-called classic theory of landscape development was formulated with reference to humid midlatitude regions, where most of the researchers concerned still live at the present day. The challenge to the classic theory came from sub-humid Africa. The collision of opposing views proved, at times, bitter. It involved some matters of high irony. Central to the dispute is the way in which hillslopes develop through time.

For purposes of the historical view, we may concede that, with the uplift of new land, there begins an orderly sequence of sculptural changes in the new landscape. To this sequence is given the name *cycle of erosion.* Since the most familiar parts of the world's land-area are being modelled by running water – rain and rivers – erosion by running water is called *normal erosion*, and the erosion cycle effected by running water is called the *normal cycle.* A sour critic might suggest that erosion by ice or wind ought to be classed as abnormal erosion: but such a critic would risk being thought captious.

The concept of an erosion cycle ultimately owes much to Darwin, for it is essentially a concept of evolutionary development. The idea of natural selection applies directly to the competition among rivers. In this competition, it is only the fitter which survive. The weaker are at the least dismembered to some extent, at most consumed altogether. Small wonder, then, that at about the turn from the nineteenth century into the twentieth, the notion of evolving landscapes proved gripping.

As systematized by W. M. Davis in North America, and as widely accepted in the English-speaking world, the normal cycle is divided into three stages. These are given the forceful metaphorical names of youth, maturity, and old age.

Youth begins with the uplift of new land. Uplift is usually taken to be rapid, so that rivers formed on the emergent surface have little scope, at first, to deepen their valleys. The cutting of valleys continues after uplift has ceased. That is to say, the later development of the landscape is under the control of the downcutting streams. The sequence involved can be very briefly summarized. So long as portions of the newly uplifted surface remain intact, the landscape is still young (Fig. 34, *a* : 1). Landscape maturity sets in when the original uplifted surface is consumed – that is, when opposing valley-walls meet at the crests of divides (Fig. 34, *a* : 2). Maturity merges into senility by an imperceptible progression. However, senility is typified by the progressive subduing of the general relief, as hillside slopes are reduced and as upstanding hills and mountains are worn down (Fig. 34, *a* : 3–*a* : 4).

The end-product of the cycle, on this view, is the *peneplain* – a vast (but undefined) expanse of terrain with low relief, low general

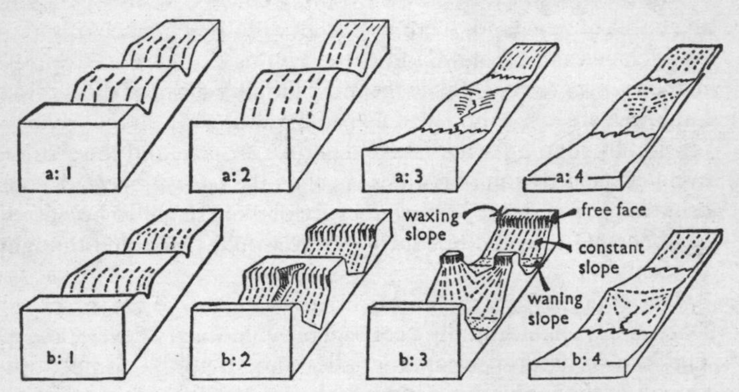


**Fig. 34 Two schemes of the normal cycle: (*a*) according to Davis; (*b*) modified after King**

elevation above the controlling base level of the sea, and subdued convexo-concave slope profiles.

In these terms, examples of young landscapes include the western borders of Mexico, the highland edge of eastern Australia, and parts of the Mediterranean coastlands (Plate 30). In these three parts of the world, broad gentle slopes remain intact at high levels between the precipitous sides of deep valleys. Rivers plunge over inaccessible and little-known falls (Plate 28). For the most part, rivers are cutting into bedrock.

To digress briefly: one process of downcutting is especially noteworthy. It consists in the drilling of potholes in channel beds. Potholes are round pits, mainly up to a few feet in diameter and several feet deep. They begin their development with the opening of hollows in the bedrock. Joint-blocks can be loosened, lifted, and removed. Once a hollow has been created, however jagged in outline, it can trap fragments of rock. And these, however angular to begin with, can be rounded into pebbles as they are swirled round by the current. In time, therefore, the potholes come to contain rounded pebbles, and themselves to be ground into rounded outlines (Plate 27). Hundreds of potholes occur on some stream reaches, the bed of the channel being drilled as if it were a giant cylinder block.

Waterfalls are often claimed as typical of a young landscape. They are liable to develop wherever a stream plunges over resistant rocks and excavates the weaker rocks beneath (Plate 28). The height of a particular fall may well increase for a time, as the plunging water carves a deep hollow at the base of the fall, and undermines the face. In midlatitude climates, where streams carry coarse sediment that can wear away bedrock, waterfalls seem doomed eventually to disappear. In the wet tropics, on the other hand, streams carry little or no coarse material, because chemical weathering breaks rock-waste down before it reaches the channels. In these circumstances, waterfalls can persist for very long periods.

By one means or another, the streams continue to cut down. They provide the fundamental controls for the development of valley-walls. In the Davisian cycle, a landscape is said to be mature when valley-walls intersect in sharp crests (Fig. 34, a : 2). The resulting combination of ridges and V-sectioned valleys is easy to illustrate from some areas of general gullying, but less readily identifiable on the grand scale. One often-quoted example is that of the folded Appalachians, where – especially in the central and southern parts – inhabited valleys are separated by high, continuous, and heavily forested ridges. Travel across the grain of the country is very hard going. The region, physically and culturally isolated, has preserved Elizabethan turns of speech and Elizabethan folk music, has developed its own regional accent and its own regional songs, is noted for the widespread distilling of illicit liquor, and was noted in former times for the waging of blood feuds.

It is only fair to add that the development of the ridge-and-valley belt of the Appalachians can be interpreted in other terms than those of maturity in the normal cycle. The matter remains one of persistent minor controversy.

Illustrations of the final stage of the normal cycle, related to existing sea-level, are difficult or perhaps even impossible to identify. They are scarcely to be expected. For at least the past few million years, the relative level of land and sea has been unstable. There has thus been no possibility of extensive planation controlled by a fixed base level. During the last one or two million years, considerable fractions of the world's lands were repeatedly covered by thick caps of ice. Their relief has been much affected – especially by thick glacial deposition in the general vicinity of the ice margins. Great portions of existing mountain belts have experienced marked uplift during the last 25 to 30 million years – about the shortest interval needed for really extensive planation.

For these reasons, landscapes which might be classified as old in the present cycle are anything but plentiful. Accordingly, the search for peneplains, or for remnants of peneplains, turned either to high levels or underground.

Some workers have identified peneplains in the deeply and smoothly eroded surfaces of ancient landmasses, such as now lie buried beneath younger sedimentary covers. Where the details of such surfaces are known, the general form often turns out to be distinctly subdued. Planation has certainly occurred. But since the rocks which now rest on the planated surfaces are typically of marine origin, it cannot be definitively proved that planation was the work of rain and rivers.

Other workers have identified as peneplain remnants the broad extents of erosional plateaus. The peneplain of one cycle provides the initial surface for the next. Possible examples include southwest England, northwest France, and northwest Spain. In all of these, subdued plateaus which truncate geological structure are being dissected by streams flowing in deep narrow valleys. In terms of the normal cycle, the existing landscapes are still young; but the plateau tops represent the senility of some earlier cycle.

Landscape maturity is thought of as merging imperceptibly into senility. A limit is set by sea-level to the downcutting of rivers, but the land between the rivers can still be attacked by weathering and erosion. Destruction of the land continues, long after the deepening of valleys has almost ceased. Divides are assumed to be progressively lowered. Rock-waste is removed from them by creep. Slopes near the tops of the divides curve over towards the lower ground, making possible the downhill transport of the rock-waste coming from above. The lower hillsides and the outer valley-floors are considered to become gentler in the downslope direction, as the calibre of creeping waste is progressively reduced and velocity of creep increases. Thus, a late-mature or senile landscape, according to David and his followers, evolves primarily by downwasting, displaying convexo-concave slope profiles (Fig. 34, *a* : 3–*a* : 4).

All this seems reasonable enough, until one looks at actual hillsides. Even then, people tend to see what they expect to see. Statements about the bold rolling country of Chalk downland, swelling hills, or the expression in the landscape of Hogarth's line of beauty (a sigmoid curve) reflect stereotypes rather than actuality. One possible cause of error is the inspection of slopes in perspective, as opposed to true profile. Another is the lack of instrumental survey. But an impartial observer can scarcely deny that, in practice, many hillside slopes are straight (Plate 31).

The angle of slope of a straight hillside varies according to rock strength. Hillsides cut in the clay formations of southern England and the Paris Basin slope at angles of 10 to 15°, while slopes developed on the limestones and sandstones of the same regions attain 25° or 30° (Fig. 36). The limiting angle in many areas is 32° or 33°. Such an angle permits transportation of rock-waste down the hillside. It is closely similar to the angle of the rest of the loose debris, and identical with the angle at which sand begins to slip on the lee side of a desert dune. The scree in Plate 3 has a slope of 33°.

Observation of straight hillsides, and measurement of slope angles, led L. C. King to propose a system for the erosion cycle which differs markedly from the system of Davis. When King made his first major pronouncement (including the redefinition of *peneplain* as *an imaginary land-form*), uproar broke out. In retrospect, and for reasons to be described, the whole circumstance was strongly charged with irony.

King's system was announced in the early 1950s, shortly after Horton had begun the revolution in the analysis of network geometry. But whereas Horton's work has little or no direct bearing on any cyclic concept, King mounted a frontal attack. Specifically, he maintained that straight hillside slopes retreat parallel to themselves – they maintain their slope angles as they are worn back

– and that concave-up footslopes grow at the expense of higher ground. The concave-up elements are called *waning slopes* or *pediments*. The straight-sloping hillsides are *constant slopes*. If a capping of resistant rock exists, its vertical edge constitutes the *free face*. At the top of the free face, the attack of weathering carves a narrow curve-over, the *waxing slope*. The combination of these slope elements, and their evolution through time, is summarized in Fig. 34*b*.

In this system, once straight profiles have been established on the valley-walls, parallel retreat ensures the consumption of the high ground. The constant slopes are themselves consumed from below, as the pediments are extended. In strict contrast to what is called for in the Davisian sequence, residual hills continue to rise sharply from the lower ground, well into the closing portion of the cycle. The end-product of the cycle according to King is the *pediplain*, a subdued erosional surface formed by the coalescence of next-neighbouring pediments.

Not all of the standard slope elements need be present in a given total profile. For instance, on the rather weak sedimentary rocks of southern England, the free face is rarely developed. The somewhat resistant cap-rocks develop a segment of constant slope, significantly steeper than the next lower segment on weaker rocks beneath, but still far from vertical (Fig. 36*a*, top right). Nor is a contrasted cap-rock necessary for the development of the free face – constant slope combination. Constant slopes can encroach on rocks which, higher in the slope profile, display free faces. Free faces can themselves be complicated by differences in rock strength. The free face in Plate 31 is, in actuality, a sequence of distinct free faces, separated from one another by narrow belts of constant slope, formed where thin shale bands occur between the massive beds of thick limestone.

Cursory inspection of the English Chalklands for purposes of comparison with King's slope model can be misleading. Waxing slopes, which are minor features of the slope system, seem on the brows of some Chalk hills to be unusually large. Exploration of the subsurface can be expected to show, in these cases, that the Chalk has been shattered, from the surface down, by frost action in a former climate. If large waxing slopes on shattered hill-brows are associated with aprons of frost-sludge at the hillfoot, then the illusion of convexo-concavity may be compelling. However, as King emphasizes, the Davisian cycle system was developed for application to temperate climates.

The bitter resentment aroused by King's interpretation of the cyclic sequence had multiple causes. In the first place, the Davisian system provided a general conceptual framework, which, as indicated, was generally relied on among English-speaking workers. Secondly, King's original field examples came from sub-humid Africa. At the time when King wrote, pediments were taken by many to be typical products of erosion in dry climates, and thus irrelevant to the landscapes of humid midlatitudes. Thirdly, King purported to base his concept of the development of hillsides on the analyses of a German geologist, Walther Penck. Finally, King sought to connect his system of the erosion cycle with the effects of continental drift, which in the early 1950s had still to return to favour.

Penck was objectionable to English-speaking workers, not perhaps so much as having written in German, but as having employed an extraordinarily difficult style. There is a partial excuse for his obscurity and ambiguity, in that he died young. The book which contains his chief pronouncements was assembled by his father, Albrecht Penck, from Walther's notes. However, there is no doubt that Penck regarded the chief aim of landscape analysis as the reconstruction of the history of geologic structure. Here was something very different from the historical studies of landscape development which the Davisian system had promoted. Also, Penck purported to explain how staircase-like flights of erosional steps could be produced as a result of continuous uplift. Few have claimed to follow his argument here. If he was correct, however, then the stepped erosional surfaces investigated by the followers of Davis need by no means all be of different ages, as they were axiomatically assumed to be.

Penck regarded himself as elaborating the Davisian system. Davis himself thought otherwise. It was Davis who represented Penck as an advocate of parallel retreat of hillside slopes. King, taking Davis to be correct in this, also credited Penck with the concept of parallel retreat, and acknowledged him accordingly when he, King, identified cases of parallel retreat in his African field areas. If parallel retreat is the norm, then backwearing prevails over downwasting.

A proper connection with Penck comes in the use of the terms *waxing*, *constant*, and *waning* for three slope elements. Penck had identified waxing, uniform, and waning phases of landscape development. Without distorting things too greatly, one can state that the early phase of the cycle can be looked on as one where waxing slopes dominate – although Penck can be read as saying that whole valley-sides are, in this phase, convex-up. The phase of uniform development, approximately represented by Fig. 34, *b* : 2, includes dominance in the landscape of constant slopes. In the waning phase, the pediments take over, decreasing in angle at given points as they develop.

The relationship of pediplanation to continental drift, as conceived by King, involved the identification for South America, South Africa, India, and Australia of parallel series of major pediplains. Correlation between continent and continent was, in King's view, evidence that the four landmasses have an identical erosional history. Identity of erosional history implied an identity of origin, as parts of the southern super-continent, Gondwanaland. The disruption of Gondwanaland King ascribed to continental drift – an idea sadly out of favour in the northern hemisphere during the 1950s.

A crucial question was – and, indeed, still remains – the climatic significance (if any) of pediments. These concave-up land-forms are widely acknowledged as occurring in deserts and semi-deserts today (Plates 78, 80). Their mode of origin is still uncertain. In fact, they may develop in different climates by different processes. King, working mainly in regions where rain, when it comes, descends in violent thunder-showers, relied on turbulent surface runoff. He pointed out that the pediment profile resembles the long-profile of a stream, and interpreted pediments accordingly as stream-beds of sensibly infinite width. Many pediment profiles are in fact semilogarithmic curves, such as can be used to describe stream profiles.

Difficulties arise, however, from the fact that numerous pediments, and all in some areas, are thickly veneered with rock-waste, which frequently consists of rolled gravel. Dissected relict pediments in the vicinity of Denver, Colorado, are capped with 10 to 20 feet of coarse alluvium. Erosive sheetflood could be expected to strip pediments bare. Some pediplains are sliced by multitudes of small channels; on others, one can travel very long distances without seeing a channel at all.

Multiple channelling, and possibly also veneering by rock-waste, could be explained by climatic change. On the other hand, we have not yet considered the climatic implications (if any) of pediments in general. If pediments can form only in dry climates, then their occurrence in regions which are now rainy gives evidence of former dry conditions. A related consideration is the relationship of pediments, especially those in Africa, to the residual hills which rise sharply from them.

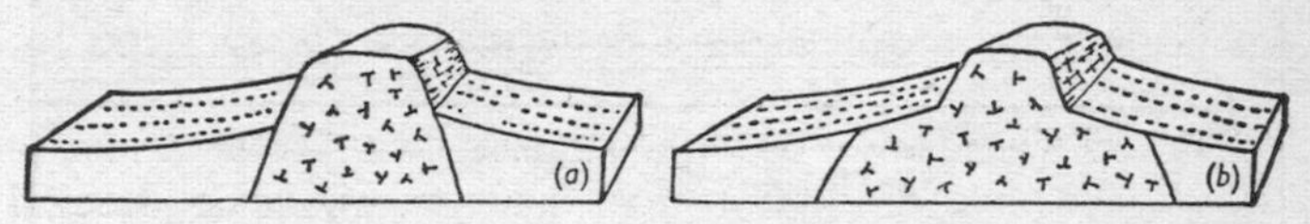


**Fig. 35** Inselbergs, with break of slope (*a*) on and (*b*) off the granite contact

Two main types of relationship exist (Fig. 35): either the pediment-head coincides with the junction between rock types of very different strength, or it does not. In the second case (Fig. 35*b*), the pediment has clearly encroached on the resistant rock. In the first case (Fig. 35*a*), there is no encroachment. Whereas in the second case the hillside may be evolving by parallel retreat, in the first case something very different is likely to occur.

Both upstanding hills in the diagram deserve the name *inselberg*, island-mountain. In a late but not final stage of its development, one would expect a pediplain to be studded with inselbergs. But where the resistant rock rises abruptly in relief, directly from the junction with weaker rock, a special type of inselberg, often develops. This is the *bornhardt*, which develops by casting off sheets. The unloading which accompanies erosion of the cover permits the opening of concentric shells of expansion-joints. Hence, bornhardts are typically dome-like in profile. Examples include Stone Mountain and similar neighbouring inselbergs near Atlanta,

Georgia. Coarsely crystalline rocks are the most common types on which bornhardts develop; but the colossal Ayers Rock in the Australian Centre, composed of sandstone, is also developing by means of sheeting.

The possible climatic significance of bornhardts is that many, if not all, appear to have originated during the destruction of a deep waste-mantle produced by weathering in hot wet climate. An extreme view, diametrically opposed to that most widely held, is that pediments also can be left over from much more humid climates than those of existing pedimented landscapes.

Work in humid middle latitudes throws much doubt on the

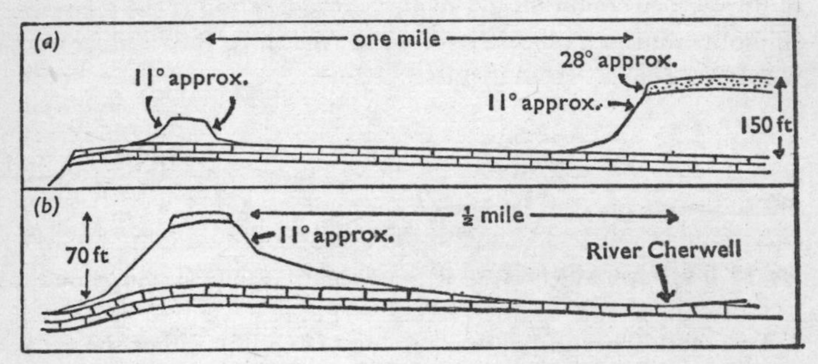


Fig. 36 Some erosional slopes in Northamptonshire

climatic significance of pediments. Fig. 36 illustrates the result of reconnaissance survey on some hillsides and valley-floors in the English Midlands. As shown in the upper diagram, the angle of the constant slope at the margin of a plateau (right) is duplicated by the constant slope of a tiny outlying hill. It would seem here that parallel retreat has occurred. However, between the main scarp and the outlying hill a formation of fairly resistant limestone has been cleanly stripped. The section lies very close to the top of a divide. The drainage has not yet succeeded in breaching the limestone.

The second section represents one side of the Cherwell river-valley. Once again, the constant slope appears. Below it, however, and cut in the same clay formation, appears a concave-up profile, that of the pediment. Numerous parallel examples could be adduced. Pediments, defined as concave-up, rock-cut features with semilogarithmic profiles, exist in cool-temperate humid climates. They can be shown to be still evolving – in certain cases, mainly by the agency of creep. Much recent work on the translocation of rock-waste, however, emphasizes the role of throughflow. This is the process whereby fine-grained material is flushed downslope through the waste-mantle. Clay particles can move even through a waste-mantle which is not particularly porous.

Pediments can develop, then, in cool-temperate humid climates. Some at least are still evolving, where they occur in regions of dry to very dry climate. Pediplains have been identified in the Amazon rain-forest, although, understandably enough, they have been used from time to time as evidence of former aridity. Workers in regions of powerful frost action recognize frost-produced pediments. We can scarcely accept pediments and pediplains, on the available evidence, as climatically diagnostic. Indeed, the weight of evidence goes to support King's contention that the pediment is the fundamental landform.

A number of investigations, both of constant slopes and of pediments in particular, and of slope systems in general, have included detailed and careful instrumental survey, plus computer analysis of the observations. The outcome is the formulation of slope models, different in all cases from the models of Davis and King, and in general highly elaborate. However, the investigations in question were, from the outset, very little directed to the conflict between the two cyclic systems which have been discussed. This conflict still remains.

A possible resolution, at least of a partial kind, comes from the finding that some slopes, at least, evolve by replacement. They are reduced at the expense of slopes of lower angle, which grow up from below. Particular cases of slope replacement are the upward extension of a constant slope, at the expense of a free face developed on the same rock body, and the encroachment of pediments on constant slopes. But really careful survey shows that, in some cases at least, the constant slope is itself segmented, without any relationship to rock type – the different segments are all cut in rock of a single kind. Segmentation of pediments is also known, but may prove to be rarer than segmentation of constant slopes. To take an example of the latter: detailed re-survey of the profiles illustrated in Fig. 36, plus additional profiles in the same general area, has shown that a constant slope generalized at, say, 11°, can actually consist of an upper segment with a slope of 12°, and a lower segment with a slope of 10°. The process of slope replacement appears among Penck's theoretical analyses.

Field investigation indicates that slope replacement is associated with abrupt changes in the strength of the waste-mantle, and especially with changes in cohesiveness. Reduction of cohesiveness reaches a threshold, where the original slope angle can no longer be sustained, and beyond which a new slope with lesser angle develops. Hence the alternative name for slope replacement, *threshold-slope decline*.

A reader is entitled to fit the concept of threshold-slope decline into the cyclic system in any way that he sees fit. On the one hand, such decline does imply a progressive reduction of hillside slopes, and in this sense accords to some extent with the Davisian thesis. On the other hand, while straight slope segments persist, they seem to be undergoing parallel retreat, and thus to indicate agreement, to some extent, with the system developed by King.

Before we can claim that a firm and general compromise is at hand, however, we need to know far more than we know now about the applicability of threshold-slope decline. It has been studied so far mainly in humid midlatitude areas. Since it appears to require the presence of a waste-mantle, if not indeed of a soil, its possible relevance to contrasted climatic conditions needs to be thoroughly tested. Conceivably, the process might be able to operate in some climates but not in others. Also, it might be able to operate on some rock types but not on others. We might guess, perhaps, that parallel retreat and threshold-slope decline constitute alternative modes of the development of hillsides. As so often, we await the production of additional basic data.

In any event, the study of landscape in cyclic terms is inevitably timebound. Adherence to the Davisian system led certain schools of researchers to concentrate on landscape history – erosional platforms, river terraces, and their implications for uplift of the land or change in the relative level of land and sea. But whereas geologists have never been worried by the inclusion of historical geology as an integral part of their science, there are geomorphologists who see landscape history as a distraction from the study of process. Process-response investigations are timeless. Moreover, they fit comfortably into a conceptual frame developed first for physics and biology, and now widely used in natural science and social science in general.

This frame is that of systems. The simplest possible definition of a system is: a set of objects or attributes, possessing internal structure, and separated from the rest of the world by a definable boundary. On first reading, this definition might seem to promise little help in the investigation of landscapes. Consider some examples, however. A glacier constitutes a system; so does a river network, a lake, the slope sequence from waxing slope to pediment and to river-bank; so also does the atmosphere or the ocean. The very fact that we give these things names indicates that we distinguish them from the rest of the world. They have objects (components) and attributes (characteristics). They have internal structure – their components are linked together, either physically or by their modes of behaviour.

Most important of all: systems can be classified as either open or closed. Closed systems receive no inputs, nor do they make outputs. An open system, on the other hand, can receive inputs and make outputs of energy, of materials, or of both. In the study of landscapes, we deal entirely with open systems.

Repeated inputs of energy can be expected on account of crustal movement. It is possible to argue that injection of energy into the erosion cycle merely brings into operation a new closed system, since the end-result must be that the energy will be dissipated as the landscape is reduced to a low-lying erosion plain. Against this, we can come to a very different view by regarding repeated inputs of energy as inevitable.

Similarly, repeated climatic change – which, like repeated crustal movement, is attested by the record – increases doubt about the usefulness of the cycle concept, however this be framed. Open-system thinking generally diverts attention away from hypotheses of ultimate planation, and towards the idea of shorter-term conditions of near-equilibrium.

For instance, a reach of river channel receives inputs at its upper end, and experiences outputs at the lower. While dimensions and form of channel vary in response to variations in discharge and sediment-load, there is usually a set of modal values, which tends to be re-established after disturbance. The system is, within limits, self-regulating: its long-term history is irrelevant. A given piece of landscape, such as a small catchment, appears to change in form rather quickly in its early history, but slowly, if at all – apart from general lowering – in subsequent times. A near-equilibrium is established, involving stream net, channel slope, channel habit, discharge, and angle of hillsides. It seems more purposeful to inquire whether or not this near-equilibrium has been attained than to ask what stage has been reached in development, and how the stream net originated.

In this way, the classification of landscapes is being released from its earlier timebound state, and is taking on an attribute of timelessness. A direct parallel is with dimensionless numbers – e.g. ratios, which are widely used in non-timebound geomorphology. The former classification of streams into young, mature, and senile is in any event due to be abandoned, because the criteria used are invalid. And the day may well come, and come soon, when the similar classification of landscapes is let to lapse.

One persistent difficulty in the investigation and discussion of slope forms and river channels is linguistic. The very use of the terms *peneplain*, *downwearing*, *backwasting*, *parallel retreat*, *grade*, and others, implies that the concepts to which these terms relate have some kind of actuality. It will probably be many years before ill-directed wrangling ceases, simply because the terms exist: the potent charges which these words carry activate a great deal of would-be technical literature, and inflame many a verbal debate. But things are nevertheless moving fast. Geomorphology is increasingly engaged in critical experiments, both in the field and in the laboratory. To its credit, it becomes progressively less descriptive and progressively more concerned with theory. If this means that geomorphology will eventually become a specialized branch of applied statistics, the change is not to be withstood.

# CHAPTER SEVEN

## *River-Profiles*

THE long-profile of a river is the two-dimensional curve shown by a section along the river from source to mouth (Fig. 37). It is often simply called the profile, unless cross-profiles – sections across the valley – are being discussed at the same time. Development of river-profiles sets a problem which has produced the most lengthy,

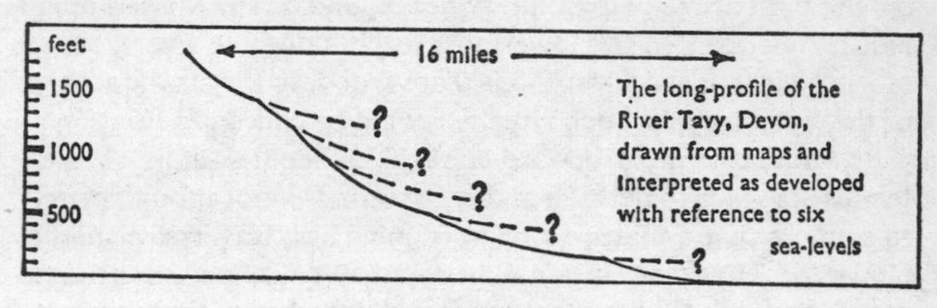


Fig. 37 A long-profile

misleading, and sterile discussions in the whole study of earth-sculpture.

Consider for a moment what is implied by the emergence of new land from beneath the sea, and its eventual reduction to a low level by the action of running water – that is, by the course of the normal cycle of erosion. The early rivers which form on the emergent land rise high above sea-level. It can safely be assumed that their gradients are, to begin with, irregular. From place to place their profiles are interrupted by falls, rapids, and lakes (Fig. 38*a*).

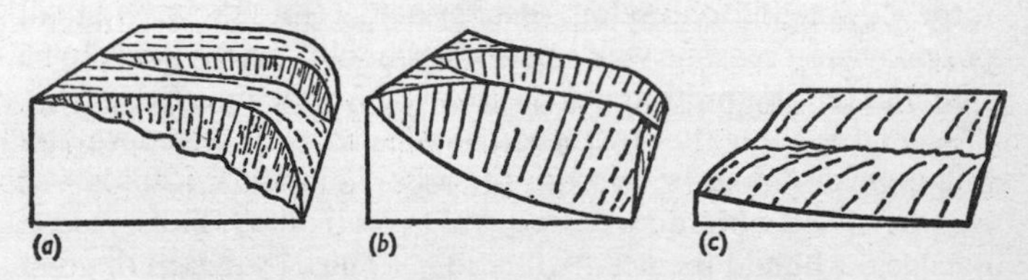


Fig. 38 Evolution of long-profile, according to Davis

In the classic scheme of the erosion cycle, such streams would be called young. The kind of average slope involved is illustrated by several main headstreams of the Amazon, which descend the flanks of the Andes through drops of 2000m in 200km – an average of 1 in 100. Many rivers on the Pacific side of Mexico descend even more steeply, with average slopes of 1 in 30.

Streams flowing across a low-lying plain, whether a plain of erosion or a plain of deposition, are certain to have low average slopes throughout. The trunk channel of the Amazon, taking 3000km to fall 200m, has an average slope of 1 in 15,000. Slopes of the same order occur on the Irtish, Lena, and Yenisei in Siberia, on the Parana-Paraguay in South America, and on the Mississippi in North America. On the lowermost reaches of these rivers, slopes of 1 in 50,000 or even 1 in 100,000 are found. In the classic scheme of the erosion cycle, such streams would be called senile.

So far all is straightforward enough. Difficulties set in when attempts are made to define a mature stream. It is not enough merely to suppose that a mature stream is one which traverses a mature landscape. However stream maturity may be defined, there is no reason to think that it must set in exactly when divides are converted into sharp-crested ridges. For instance, if the uplift which begins the cycle is a modest one, the streams may be able to readjust their profiles while parts of the uplifted surface remain intact. If on the other hand the uplift is very great, the whole of the uplifted surface might be destroyed, well before rivers had accommodated themselves to the new situation.

Many effects have been made to define mature streams in terms of their long-profiles. These efforts depend on the assumption that, given enough time, a stream will develop a smooth profile such as the one illustrated in the foreground of Fig. 38*b*. This in fact is the form which tends to develop, as many actual cases show. We shall presently meet reasons why a perfectly smooth curve is likely to be rare. Let the assumption stand, however, for the time being.

Any lakes along the stream course will eventually disappear, by filling, by draining, or by both processes combined. Rapids – at least outside the humid tropics – will be worn away. Falls – again outside the humid tropics – will also go. Thus, by means of gross adjustments, major irregularities in the profile will be smoothed out.

The removal of minor irregularities is also attributed to combined erosion and deposition. The argument runs as follows: in a reach where the stream has too little energy to transport its load, deposition occurs and the slope is steepened; in a reach where there is excess energy, this is expended on erosion of the channel bed, and the slope is decreased. Pursuit of the argument leads on to the idea of balance – balance between power to do work, and work that is to be done. As a matter of record, *work* in this context has been defined wholly in terms of the transport of sediment. A stream is supposed to develop a slope which, in any given reach, will permit the transportation of the sediment delivered from upstream or from the immediate valley-sides. A stream with its slope delicately adjusted between power and work is said to be *graded.* Its condition is the condition of *grade*. Here, perhaps, is also a definition of maturity?

Although there are roughly as many definitions of grade as there have been writers to formulate them, the above statement does no violence to the concept. A highly interesting corollary follows. Once a stream becomes graded, with power and work equalized, it loses all tendency either to cut or to fill. Prolonged and pronounced erosion of the channel bed ought not to be looked for.

Now look at what must happen in nature. Fig. 38*b* shows the profile of a graded stream. The stream has attained grade while much high ground remains in the middle part of the basin, and especially in the upper part. All the substance of the tall divides must be carried down the channel of this and other streams, before the landscape can be reduced to the subdued and low-lying form illustrated in Fig. 38*c*.

In these several ways, the concept of grade proves to be elusive, not to say possibly confusing. It appears unhelpful in the attempt (supposing such an attempt to be made) to define stream maturity. Other attempts rely on the plan geometry of channels. It has been claimed that a river no longer capable of vertical erosion expends its excess energy by cutting laterally, and that lateral cutting with the indicated cause produces meanders. Thus, meandering becomes a criterion of stream maturity. How the river perceives, and makes, the choice between vertical and lateral cutting is never made clear. Nor is it explained why, if the river possesses excess energy, this is not expended on erosion of the channel bed. This pointless theme need scarcely be pursued further. All that needs to be said is that channel habit is no indicator of stage of stream development.

One purported explanation of the concave-up form of the profile relies on the stream source and the mouth at sea-level as fixed points, between which the profile is variably lowered. Erosive power is said to be small in the extreme headwaters, where streams are tiny. It is also said to be small in the far downstream reaches where the stream is approaching sea-level and has a gentle slope. This leaves the middle reaches, traversed by a large channel but still well above sea-level, in the belt of maximum cutting. Hence the concave-up form.

That particular argument has become entangled with ideas about erosion in relation to stream velocity. Velocity is held to be low in downstream reaches with very gentle slopes. There is some lack of logic here, for one would expect velocity to be high on headwater reaches, where slopes are greatest. Rapid erosion might also be expected there as a consequence. On first reading, there is a certain attraction in the description of a senile stream as crawling sluggishly down the very low slopes of an erosional plain. The parallel reading for the headwaters is one of young streams dashing impetuously down the mountainside. It seems self-evident that velocities will be greatest where slopes are steepest. But self-evidence proves highly misleading.

We must rely on records made on actual streams. These records show that average velocity *increases* downstream, despite the reduction in slope. The increase is especially well documented for the Yellowstone–Mississippi system. There, a slope of 1 in 50 in the far headwaters is associated with an average velocity of about 0·1m/sec., while a slope of 1 in 20,000 near the sea is associated with an average velocity of about 0·4m/sec. (Fig. 39).

Average conditions, though, are of little interest. They were subjected to analysis in the early stages of investigation, simply because they had been abundantly recorded. What we need is analysis for conditions when the channel is full or overflowing. On large rivers, a whole series of velocity measurements must be taken, at intervals across the stream and at intervals of depth, in order to

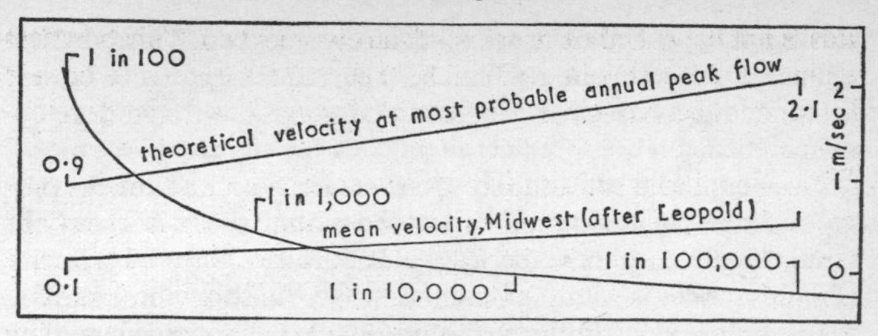


Fig. 39 Gradient and velocity

determine the mean velocity through the cross-section. Velocity can vary greatly at any given section, not only with the amount of discharge, but also in response to temporary scour or fill. Moreover, readings on large rivers are normally taken from bridges, where piers and abutments disturb the natural relationships of flow to discharge. Fortunately, enough material exists to indicate for some rivers the way in which velocity varies along the length of the channel, when this is full to the brim. Here is the point of reference: flow at bankfull stage.

Mean velocity through the cross-section increases from low-flow stages up to the bankfull stage. Velocity *within the channel* seems to increase little or not at all, when the river spills out of the channel at flood (= overbank) stages. The variation in velocity at the bankfull stage, from one reach to another, may however be so great that no general relationship between velocity and distance along the channel can be made out.

Furthermore, some rivers on some reaches are lowering their profiles by cutting into their channel beds. Other rivers on some reaches are increasing their slopes by deposition. It then becomes impossible to apply the criterion of bankfull flow in analysing stream behaviour. Fortunately, theoretical analysis comes to our aid. It provides an alternative point of reference; and it defines how velocity can be expected to change along the length of a stream. The alternative point of reference is a statistical one. The flow which shapes the channel is found to be flow at the most probable annual peak. On a channel which is neither cutting nor filling, this flow would fill the channel to banktop. But use of the most probable value permits the use of data on rivers which are cutting or filling. The flow in question can be expected to be equalled or exceeded, on the average, once a year.

Theoretical analysis predicts a tendency for velocity at the most probable annual peak discharge to *increase* slowly in the downstream direction. Reference to data on actual rivers in areas of humid climate permits numerical values of velocity to be written in, against selected values of channel slope. As shown in Fig. 39, the theoretical velocity through a water-filled channel, where the slope is 1 in 100, is less than 1m/sec. The stream here is small – its width is about 3m and its channel-full discharge about $1m^3$/sec. Far downstream, where slope has decreased to 1 in 100,000, velocity rises to more than 2m/sec. Here the stream is of the order of 1700m wide, with a bankfull discharge of $100{,}000m^3$/sec.

As indicated above, channel-full velocity on actual rivers can vary much, both between river and river and between reach and reach on the same river. It seems uncommon, though, even on very steep slopes, to record velocities of more than 3m/sec.; and some of the highest known velocities belong to large rivers near the sea, where slopes are very gentle. Such is the case on the lowermost Amazon.

The reason why slope tends to decrease in the downstream direction is that rivers – at least in humid climates – become larger in that same direction. They are progressively augmented by tributaries. The amount of flow when the channel is full depends chiefly on drainage area. The farther downstream on the trunk channel, the larger the area drained, the greater the discharge, and the larger the channel. A large channel is more efficient than a small channel, simply because it is large. As channel size increases, the average friction exerted on the flowing water is decreased. Therefore, large channels require lesser slopes than do smaller channels. For the theoretical river in Fig. 39, friction is about ten times greater at the upper end than at the lower.

To generalize: if a stream increased steadily in volume from source to mouth, its gradient would decrease steadily in the same direction – *but only if all adjustments to changes in volume were made in terms of gradient, and only if changes in gradient were controlled exclusively by changes in volume.*

Here is the crux of the problem of long-profiles. A profile is drawn in two dimensions, but the problem is a three-dimensional one. Cross-sectional size is not the only control channel efficiency. Cross-sectional shape is also important.

The most efficient channel has a semicircular cross-section. In nature, semicircular cross-sections do not occur. Channels shaped in alluvium are wider and shallower than a semicircle. Channels shaped in bedrock may be either wide and shallow or narrow and deep. In either case they are liable to be inefficient. Very wide and shallow channels in alluvium are also inefficient, the extreme situation being represented by braided channels (Ch. 8). For a given discharge at the most probable annual peak flow, channel slope is likely to be greater on an inefficient channel than on a more efficient one. The river appears to respond to increased friction by steepening its slope.

A case in point is that of Brandywine Creek in Pennsylvania. Field study has shown that this river is capable of transporting, at the bankfull stage, all the load of solid rock-waste which is fed into it. The study included the use of special sediment-traps, the recording of calibre and amount of sediment load, of transport in progress, and of course of discharge. The finding is that the Brandywine has reached a state of short-term balance between erosion and deposition. If we wish to use the term, it can be called a graded stream.

But its long-profile is by no means a smooth, concave-up curve (Fig. 40). Adjustments to variations in load, discharge, and efficiency have been made in three dimensions. The size, shape, and slope of the channel have all been modified over quite short distances. Whatever the ultimate causes of the irregularities in the profile, these irregularities seem unrelated to changes in sea-level or to movements of the earth's crust. The intervention of the fall zone separates the illustrated portion of Brandywine Creek from the control of existing sea-level.

It seems just, therefore, to regard the ideal, smoothly-curved long-profile as one end-member of a continuous series. At the other end comes the highly irregular profile of a stream which has so far had little time to remove even the gross irregularities resulting from considerable uplift. Breaks in the profile are called *nickpoints*.

There is a major nickpoint on the West Branch of Brandywine Creek, at about 550 feet above sea-level. A less strongly marked nickpoint occurs on the East Branch, at about 450 feet above sea-level. If these breaks of slope reflect changes in the relative level of land and sea, such as are suggested in Figs. 37 and 42, then they are *cyclic nickpoints*. But if they merely represent responses to change in channel efficiency along the length of the stream, then they are *non-cyclic nickpoints*. There is no ready means of sorting out one kind of nickpoint from another. As will be seen, the distinction can bear importantly on the history of landscape.

Once it is admitted that irregularities of gradient may be com-

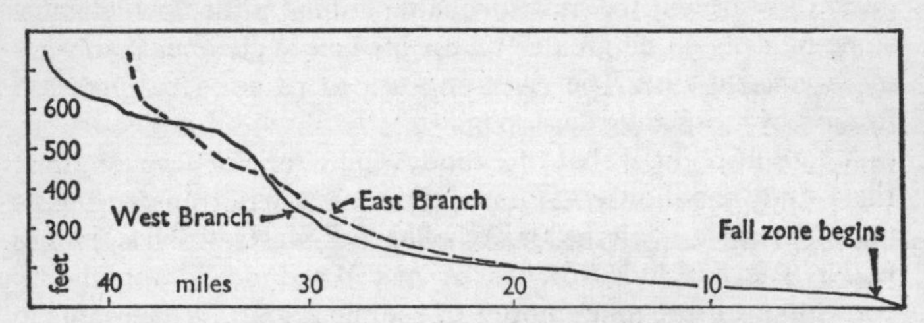


**Fig. 40 Long-profile of Brandywine Creek, Pennsylvania (re-drawn after Wolman)**

pensated by changes in the cross-section of the channel, suspicion falls on a number of venerable ideas. Among these is the view that long-profiles can be described by mathematical equations. The partial profiles recorded for actual streams broadly resemble the form of a logarithmic curve. Variations can be made in the equa-

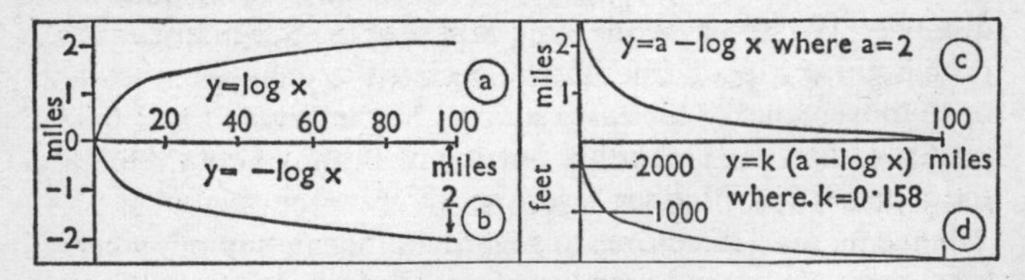


**Fig. 41 Equation for long-profile**

tions to ensure that the calculated curves come the right way up, tend towards sea-level at their lower ends, and start at the right height above sea-level at the sources. For these purposes constants are introduced into the equations (Fig. 41).

Numbers of people (including the writer) have indulged in the pastime of fitting mathematical curves to profiles plotted for actual streams. The pastime is more than a pleasant mental exercise. Granted certain assumptions, it could provide the means of relating partial profiles to the former sea-levels which controlled their development.

Profiles are developed with reference to a base-level of erosion, which may be the surface-level of a lake, the confluence with a trunk stream, or the level of the sea. In so far as a smoothly curved profile comes into being, its development is ultimately under the control of the base-level from which the profile is extended headwards. If a river with a smoothly curved profile is affected by a fall of sea-level, a new profile extends itself upstream, intersecting the remaining part of the old profile at an angle until the whole of the old profile has been destroyed. If a vanished part of an old profile could be reconstructed, the former height of sea-level could be ascertained.

Profiles can, of course, be drawn for the whole of a river-system, and not merely for the trunk stream (Fig. 42). Breaks of profile

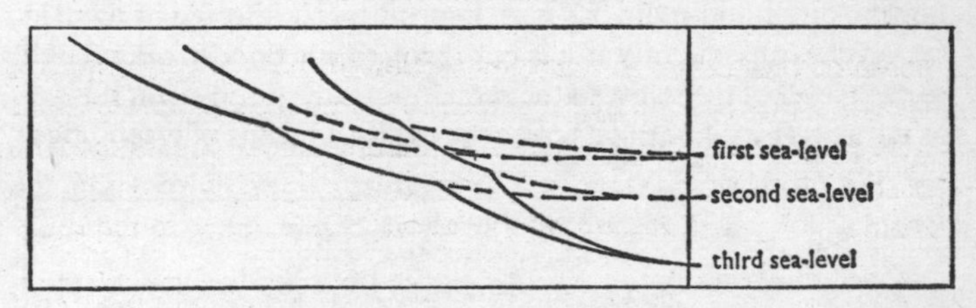


Fig. 42 Possible relation of profiles to sea-levels

often seem to occur in groups, especially if the average gradient is steep and if the diagram is drawn from maps instead of from detailed survey. Each group of partial profiles, it may be presumed, belongs to a single former base-level. Indeed, it may be easy to extend the curves by drawing, without any help from equations. But drawing involves a certain amount of guesswork, and it would obviously be more satisfactory to rely on calculation than on draughtmanship. This is especially true when little remains of the older profiles.

Unfortunately there are serious obstacles to the extension of profiles, either by drawing or by calculation. One has already been made clear in the previous discussion of the form of profiles in general – it has been shown that a regular curve may never develop in actuality. Even if it did, extension of its remains would not be easy. There is a well-known case of two elaborate equations, each giving a perfect fit to a partial profile: the former base-levels predicted by the two curves differed in height by no less than 50 feet.

Although the foregoing paragraphs have emphasized the uncertainties which afflict the reconstructor of profiles, the general principle remains valid that all rivers which reach the sea have been affected by changes in sea-level. Some rivers have been further affected by movements of the crust of the earth. Efforts to fix the heights of old base-levels are, in consequence, directed to a reasonable end. It is the practice which is difficult.

Much help in reconstruction is to be had from river terraces. Many of these consist of patches of sand, gravel, silt, or clay, standing at varying heights above the river and representing the deposits laid down on some former valley-floor. At the simplest, a river flowing on a flood-plain has been intermittently affected by a fall in sea-level. Consequently it has cut through the flood-plain related to each level of the sea, and the remains of each flood-plain survive as terraces (Fig. 43). Each terrace provides a means of extending a

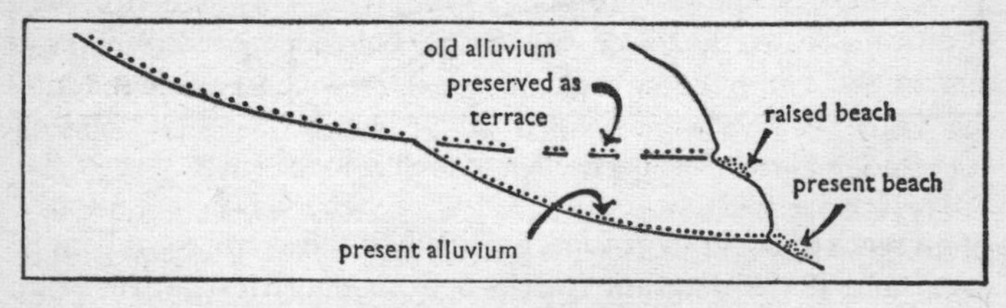


Fig. 43 Profile, terrace, and raised beach

partial profile downstream from a break of slope. In exceptional cases a terrace may continue right to the former coastline, merging into a raised beach. But the older terraces are usually much eroded, and there may be no trace of the associated former beaches on the present coast.

In the lower valleys of southern Britain the oldest terrace is at the top of the series, the next oldest is the next highest, and so on. The arrangement results from an intermittent fall in sea-level. But the observations made on the ground have yet to merge in a simple general picture, for two reasons. Firstly, it is not easy to relate the sequence of terraces established for any one river to the sequence established for any other; secondly, the number of terraces recognized in any one valley tends to increase, as field-work is carried out in increasing detail.

Interesting complications occur when rivers have been influenced in their lower reaches by changes of sea-level and in their upper reaches by simultaneous changes of climate. Very many rivers of western Europe were so affected during the Ice Age. The Ice Age did not amount merely to a single waxing and waning of the great ice-sheets. The enormous glaciers spread and decayed four times (p. 169), and in each of the four episodes there was a fall of sea-level – due to the locking-up of water in the form of land-ice – followed by a rise in sea-level when the ice melted again. Rivers cut deeply into the seaward ends of their valleys when sea-level fell, and deposited sediment in the excavations when it rose. Meanwhile, upper valleys invaded by ice or subjected to severe climate received great quantities of rock-waste – either as ice-deposits, as outwash, or as thawed-out debris which moved downhill over frozen subsoil. Thus the upper valleys were being choked with rock-waste precisely at the times when the lower valleys were being deepened. When the climate once more became genial, the deposits previously fed into the upper valleys were eroded at the same time that infilling occurred near the river-mouths.

This is not the whole of the story. In addition to the falls and rises of sea-level caused by the growth and decay of the ice-sheets, there were general falls of sea-level, well authenticated although of unknown origin. In consequence, the sea which rose on receiving glacial meltwater failed to reach the height previously recorded. The kind of effect produced by this combination of causes is shown in Fig. 44.

In a simplified way, this diagram represents sequence of changes in the profile of a single river during one of the four glacial episodes. The highest sea-level, marked *pre-glacial*, controlled the development of the gently sloping long-profile 1–1. When sea-level was depressed during glacial maximum, the lower end of the valley was excavated and the upper part infilled, in the manner just described, and the profile 2–2 was developed. Its gradient is steeper than that of the profile 1–1, for there was no time for the material fed into the upper reaches to be carried away, let alone for general downcutting throughout the course. Furthermore, the great bulk of loose rock-waste in the upper valley could not be carried down a gentle gradient like that of the pre-glacial valley. When the ice melted and the level of the sea rose once more, another profile of the pre-glacial kind developed; but because the sea now stood below the pre-glacial level, this third profile (3–3) lies below the pre-glacial profile 1–1 in all parts of the valley.

If we now use the diagram to construct cross-sections of the valley, on which the various deposits are recorded, we shall perceive some of the complexities of the study of terrace material. At the seaward end of the valley, the profile 1–1 is recorded by a terrace which stands above the present level of the river. The profile

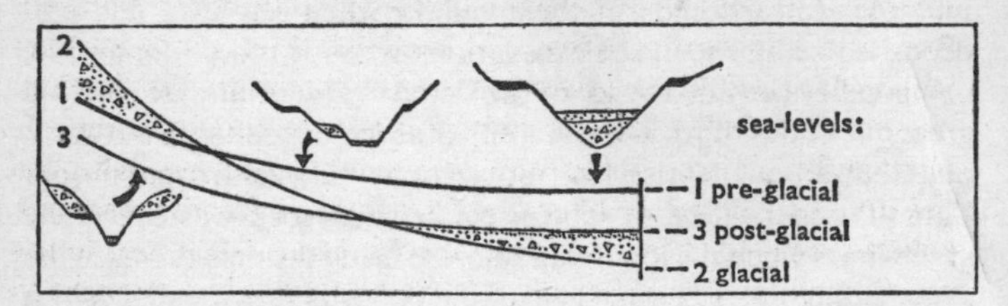


**Fig. 44 Profiles controlled by changes of sea-level and of climate (adapted from Baulig, with amplifications)**

2–2 is represented by the bottom of a buried channel, such as that proved beneath the Thames at London, beneath the mouth of the Exe, and beneath numerous other estuaries. It seems likely that numbers of such channels were developed in relation to a sea-level about 200 feet below the present level. The infilling of the buried channel began when climate was still very cold, so that the fossils contained in the lower part of the fill are of a so-called cold type.

Progressively warmer conditions are indicated by the higher parts of the sequence, culminating in the present flood-plain 3 with its post-glacial forms of life.

In the upper reaches, the space-relations of the various deposits and erosional forms are of a different sort. The pre-glacial valley floor, 1, is buried beneath deposits formed in a cold climate. If any sequence of change can be made out, it is from early-glacial conditions at the base of the fill to high-glacial conditions at the top – in other words, the sequence of this fill is one of increasing cold. Deposits of this kind are present in the headwater valleys of the Thames and Moselle. There has not yet been time for all the deposits to be removed by post-glacial streams; but even so, these streams have cut through the sludge-gravels in places and are now working in solid rock, below the level of downcutting reached in pre-glacial times.

Thus it follows that the remains of old profiles cannot always be identified merely by their height above the existing rivers. The profile 2–2 in Fig. 44 crosses the profiles 1–1 and 3–3 on the vertical plan, standing higher than both of the others in the upper valley, below both in the lower valley, and between them in the middle reaches. The possible complexities which result not from one but from four glacials can, perhaps, be imagined. The record becomes still more varied when a river has failed to complete the development of the profile related to any one sea-level, when adjustments have been made in cross-section as well as in gradient, and when the basin has been disturbed by crustal movement. But perhaps enough has been said to show, in a general way, what is involved in the practical study of profiles, and this discussion may be brought to an end with a few additional references to actual cases.

The original diagram on which Fig. 44 is based was drawn for the Rhône, which is still fed with glacial meltwater today. The mouth of the river is on the Mediterranean, far beyond the farthest limit reached by the Alpine glaciers at their maximum extent. At glacial maximum the Mediterranean stood lower than it does today, and the Rhône and its tributary the Durance laid down coarse gravels in a delta related to the low sea-level – the stony district called the Crau. The deposits of the Crau plunge beneath the present delta which is related to the post-glacial level of the Mediterranean. The present delta is a wind-swept, isolated region of marsh and lagoon, with lines of dunes marking the positions of old shores.

The lower valley of the Mississippi, having been well studied, has had a complex history of cutting and filling. Terraces mark the positions of old flood-plains. Towards the mouth of the river they slope downwards, even passing beneath the present alluvium, for this part of the crust has been intermittently depressed by the weight of the deltaic deposits, and, here again, it is possible for terraces to cross one another in the vertical plane.

Similar conditions obtain in the Rhine delta. It is not yet known whether the floor of the southern North Sea is subsiding merely because of the weight of the Rhine sediments, or whether the mouth of the Rhine has been fixed in position by subsidence. But deposition at and near the mouth of the river has been capable of promoting subsidence, with the result that deposits which lie at 400 feet above sea-level north of London are continued at lower altitudes across eastern Essex, and corresponding beds in Holland lie hundreds of feet below the level of the present sea.

CHAPTER EIGHT

# *River-Traces*

SOME of the most exciting and important problems of earth-sculpture are set by the patterns made by rivers. Certain distinctive kinds of pattern displayed by whole river-systems have already been illustrated and discussed. The degree of adjustment to structure, as indicated by the pattern of the whole river-system, epitomizes the whole history of the river since it was first initiated. Another kind of pattern – that made by the channel – shows what the river is doing at the present time.

River-traces, or channel-patterns, often assume forms as distinctive as those of river-systems. But they need close study; whereas a map of no great accuracy will succeed in representing an array of streams, channels must be examined with the help of accurate large-scale maps, on the ground, or in the laboratory.

Many valleys run more or less straight for long distances, but straight channels are not very common. A river of any length is most unlikely to pursue a straight trace throughout the whole of its course, although some large rivers include straight reaches, and small streams can be entirely straight. Generally speaking, it is fair to say that truly straight channels are confined to places where special conditions apply.

Such channels can occur on very steep slopes – for example, on the sides of glacier-troughs, where streams in little gullies flow down precipitous hillsides (compare Plate 53). Gravity succeeds in guiding these streams directly downhill. Even so, it is not uncommon to find the larger gullies developed on the lines of joints, where the rocks are cracked, or on the lines of faults, where the crust is fissured. Joints and faults can also guide streams of lower gradient in straight courses, which are, however, inset in traces of other kinds. Where dykes of rock, injected into straight fissures in a molten state, have solidified and rotted, the resulting narrow belt of weakness is likely to be picked out by running water; but an eroded dyke frequently gives room for a stream to swing from side to side in a narrow trench, so that its trace cannot justly be called straight.

In contrast to straight channels, winding channels are very common indeed (Plate 33). These belong to meandering streams, each bold curve in the channel constituting a single meander. Meanders are seldom quite regular either in shape or in dimensions; at the same time, they tend in any given reach to be much of a size and to be spaced with rough equidistance. Since meandering traces are so frequently encountered, it can be inferred that the conditions which produce them are widespread in nature. This general observation does not help in examining the origin of meanders, but at least it states a general principle which may usefully be kept in mind.

A third class of trace is that made by the braided channel (Plate 38). In cross-section a braided channel is very wide and shallow. Its ratio of width to depth is typically far greater than the ratio for single channels, whether straight or meandering. The most obvious characteristic of a braided channel is that, at stages of low flow, minor channels divide and reunite round bars of loose bed material – usually gravel, although similar forms can be built of sand. Each bar constitutes a braid. At stages of high flow, parts or all of the braids are submerged. But since the total duration of high flows is much less than that of low flows, the characteristic appearance of a braided channel is that shown in the photograph.

Although innumerable variations of trace occur on natural streams, it will be enough for the present purpose if some light can be thrown on the three classes of trace which have been specified.

Most of the subsequent discussion will be devoted to meanders. There are several reasons for the intended manner of treatment. More is known about meanders than about braids; an understanding of meanders will be needed on a later occasion; and certain mistaken ideas about meanders need to be refuted. The general import of what follows will be that straight traces belong to the early phase of stream-development on steep slopes, and that braided and meandering traces are alternative end-products of natural processes.

As a result of recent investigations, it is possible to say how meanders originate – but not why. It is well to begin by stressing that obstacles in the channel have nothing whatever to do with the initiation of meanders. This fact cannot be too strongly emphasized. It was noted more than half a century ago by W. M. Davis, but the contrary – and mistaken – view has crept back, appearing not only in the writings of geographers and geologists, but also in those of civil engineers. The error is so widespread that it needs more than a plain contradiction.

Many engineers and geologists, in discussing the origin of meanders, appeal to the mechanism of bank-caving – i.e. to the collapse of undercut banks. Geographers seem to fix their attention on small streams and on obstruction by falling trees. But banks are undercut, just as trees are undermined, *because the stream is already meandering* (Plate 32). It is quite wrong to interpret a result as a cause. For the cause of meandering we must look elsewhere.

Now although it is impossible to conceive of a natural stream with no obstacles in its channel, some natural channels are in fact very little obstructed, and artificial streams can easily be made entirely free from obstacles. It is possible, therefore, to compare the behaviour of an artificial stream in perfect conditions with the behaviour of a natural stream in conditions not far from perfect, and to contrast both with the behaviour of a natural stream which is notably obstructed.

Very lengthy series of observations have been made on the Mississippi, by members of the Mississippi River Commission and other bodies. In the lowermost 600 miles of its valley, the Mississippi flows in looping meanders on the flat bottom of a silt-lined valley. Numerous borings show that the river, far from being in contact with the solid rocks in which the valley is cut, flows over thick layers of alluvial deposits (Fig. 45*a*). Its meanders do not encounter the solid rocks; nevertheless, obstacles are present, From time to time one of the shifting loops impinges on the fill of an abandoned channel, where the plug of clayey sediment is more coherent than the flood-plain alluvium generally. The coherent fills of clay, resistant to erosion, check the development of meanders and cause them to be distorted. No clearer demonstration could be needed that obstacles are adverse – not favourable – to meandering.

Remarkable instances of distorted meanders come from the Appalachians, where several streams, traversing the Martinsburg Shale formation, have developed meanders which are greatly elongated in the sideways direction. Strahler, discussing the Conodoguinet – a tributary of the Susquehanna – concludes that the elongation is a response to the influence of slaty cleavage. The axis of the meandering Conodoguinet runs approximately parallel to the lines of slaty cleavage. On the outsides of bends, the sides of the tight layers of rock are exposed, whereas in other positions

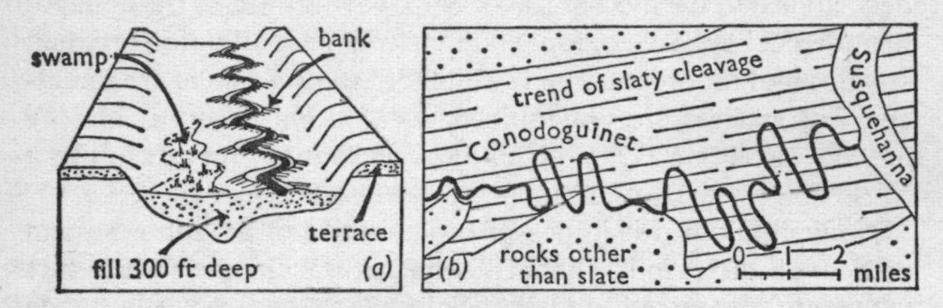


Fig. 45 (*a*) The Mississippi flood-plain (diagrammatic)
(*b*) Distorted meanders of the Conodoguinet (adapted from Strahler)

the stream is working against the edges of a slab of slate. Consequently, erosion is greatly facilitated at the extremities of the loops. During incision to depths of 100 or 200 feet, the river has extended its loops sideways to distances ranging from 450 to 900 yards (Fig. 45*b*).

Although the width of the meander-belt has thus been greatly increased, the average wavelength of meanders has been little affected. The value of wavelength in distorted reaches very closely resembles that in unaffected reaches. The average wavelength of some 4000 feet is little more than 15 times the width – 260 feet – of the stream-bed. Not only the Conodoguinet, but also the North and South Forks of the Shenandoah, the Middle river, and the Conococheague, clearly show that meander-development is checked or disturbed by obstacles in the course of stream.

Meandering in the absence of obstructions is well illustrated by the experiments on laboratory streams performed at Imperial College, London. A stream-trough was lined with granulated perspex, in order to simulate sandy alluvium in a natural valley. Into a straight and symmetrical channel down the centre of the trough, water was fed at a constant rate. After no great length of time, shoals appeared at intervals on the bed of the channel. Winding round the shoals, the stream began to cut into the banks, alternately on the right and on the left. Meanders had been produced – in the absence of obstacles, in uniform material, and with constant discharge.

It was concluded from these experiments that meanders develop in two phases. In the first phase, alternating hollows and shallows appear in the stream-bed, controlling the direction of the current. This phase is short-lived in experimental streams. As soon as the stream begins to swing from side to side, the form of the banks affects the flow of the current, and through it the form of the bed. These are the conditions of the second phase of development, which is represented by natural meandering streams. Once meandering, a stream continues to meander unless its windings are seriously obstructed by obstacles, or unless it is converted to some other kind of trace.

Observations made on natural streams, notably by the U.S. Geological Survey, agree completely with the experimental results summarized above. Leopold and his fellow-workers find that hollows occur in the beds of straight streams, that the hollows are spaced with rough equality, and that their spacing corresponds to the spacing expectable from meanders. It seems difficult to exaggerate the significance of these findings, which seem to illustrate the first phase in the development of meanders – namely, the alternate shoaling and deepening which precedes side-to-side swing.

Marked contrasts can occur between one river and another, in respect of boldness of meander curvature. Nevertheless, meanders display a strong family resemblance, wherever they are found. This is partly on account of the relationship of channel width to meander wavelength.

As for long-profiles, discharge at the bankfull stage (or at the most probable annual peak flow) provides a basis of reference – in particular, for the measurement of channel width. Width is taken in completely simple cases between the tops of the channel banks. Theory predicts that width should vary with the 0·55 power of discharge, as defined above. Observation agrees. Theory suggests that wavelength should be some multiple of bed-width. To an extent, observation again agrees. Wavelength is commonly found to be between 8 and 12 times width. A likely central value is 11 times.

If wavelength is 10 times width, then the hollows (pools) at meander bends will be spaced at 5 times width. Pools on roughly straight channels are often found to be spaced at 5 to 6 times width. Hence the inference that they belong early in the sequence of meander formation. But some channels seem to be arrested here. They have developed sequences of pools and shallows (riffles), but show no tendency to go further and develop side-to-side swing. Here is an outstanding puzzle.

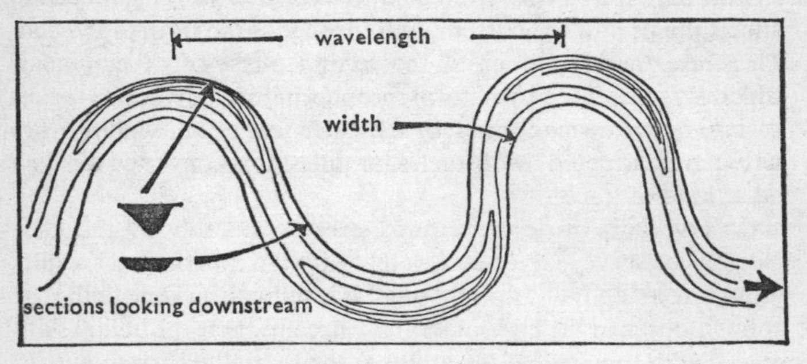


Fig. 46 Width, wavelength, and bed-form of a meandering stream

The reason for the rather narrow range of wavelength/width ratio lies in the response of the flowing water to channel curvature. Resistance to flow round the bend is least, when radius of curvature is between 2 and 3 bed-widths. Numerous meanders are highly distorted – as by the obstacles provided by unusually resistant material in the banks. But wherever possible, the river will tend to adjust its bends to the standard form. The increased resistance imposed by an unusually tight or unusually open bend will be taken up in bank erosion. This in turn will reshape the bend towards the plan illustrated in Fig. 46. We can state as a general conclusion, that the meander pattern enables a river to do the least work in turning.

Analysis can be made to show that the typical meander plan is a most probable shape. It also demonstrates an apparent paradox –

that meandering actually minimizes changes in channel direction. A necessary assumption here is that the channel should be other than straight. We can thus, if we wish, conclude that the observed pattern of a meandering channel results from random variation. We can certainly conclude that a meandering channel is efficient. We can suspect that a straight channel tends to be unstable.

Development of meanders does not cease with the appearance of swinging loops. Each loop tends to enlarge itself by eroding the bank on the outside of the stream-curve. Even where bank-erosion is rapid, however, there seems to be a limit to the process of enlargement.

Some meander-trains seem to have reached a state of stability, in which the loops have little tendency to grow larger. With other trains, continuing growth brings into operation the process of cut-off, whereby the loops are short-circuited and their growth automatically checked. Erosion of the outer banks narrows the necks of land between successive loops (Fig. 47*a*). Bank-erosion is likely to be most severe on the downstream sides of bends, which overtake

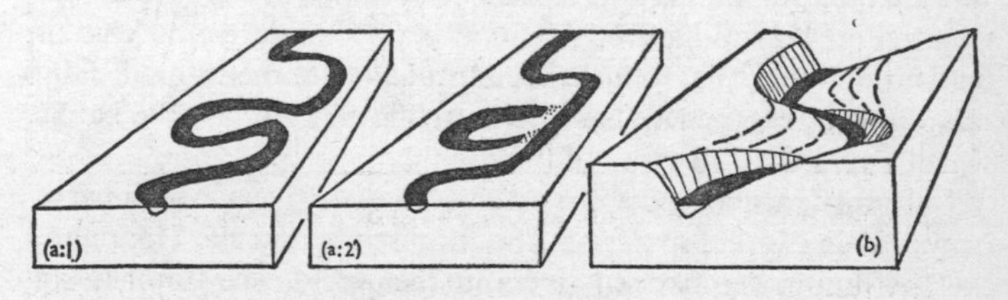


Fig. 47 (*a*) Cut-off
(*b*) Ingrown meanders

the upstream banks next below. In time of flood the diminishing necks can be broken through. A cut-off loop does not long remain in contact with the new channel, for its ends are soon blocked by sediment. The horseshoe-shaped lake thus sealed off is doomed eventually to be filled in by flood-laid rock-waste and by the debris of plants.

Individual distortions to which meanders are subject, and the frequent short-circuiting of single loops, are variations on the general theme of down-valley shift. Trains of meanders migrate downstream, so that the whole floor of the valley is worked over by the river. This is why many meandering streams flow through a strip of alluvium, which, since it is flooded from time to time, is known as a flood-plain.

The topic of the origin of flood-plains brings us back once more to the matter of discharge. It was stated above that the form of the channel is determined by the discharge at bankfull. At lower stages, loose material is rolled from the shoals into the hollows. At bankfull discharge, the hollows are swept clean and the shoals are built up. At higher stages – i.e. at overbank discharge – the channel can no longer contain all the water that comes down the valley, and the flood-plain is inundated by a sheet of muddy water.

In some valleys a great deal of sediment settles from the spreading flood-water. Deposition is concentrated near the edge of the main channel, where the speed of flow is rapidly decreased. Wide natural embankments are formed, sloping gently outwards from the edge of the main channel towards the sides of the valley. In extreme cases the outer parts of the flood-plain become seriously waterlogged and marshy, as in the backswamps of the Mississippi valley (Fig. 45*a*). Wherever the flood-plain is under some kind of cultivation, it is usual to find the natural embankments artificially raised and strengthened. Natural or artificial, these are the banks which rivers burst in times of flood.

Naturally raised banks are associated with rivers which carry heavy loads of solid material. The Mississippi and the Hoangho, the best-known embanked rivers in the world, are notoriously charged with sandy particles. On a very much smaller scale, the formation of embankments is illustrated by the little stream which, discharged from a gravel-washing plant, also transports a great deal of debris in proportion to its size (Plate 35). But some natural rivers have no raised natural banks, despite the fact that they traverse flood-plains and overflow from time to time. It must be concluded, therefore, that the alluvium of their flood-plains has not been deposited mainly by flood-water. If it had been, the flood-plains should be slightly convex upwards, and not flat as they are in actuality.

Flat flood-plains are the products of migrating meanders. Erosion on the outside of a meander is accompanied by deposition on the inside. Tests made with identifiable material show that debris removed from the outer bank comes to rest – for a time – on the inside of the next bend downstream: that is, on the same side of the channel. As the meanders shift down-valley, the crescents of alluvium on the inner sides of the bends grow correspondingly (Fig. 48). In time the deposits coalesce over the whole valley-floor, in a

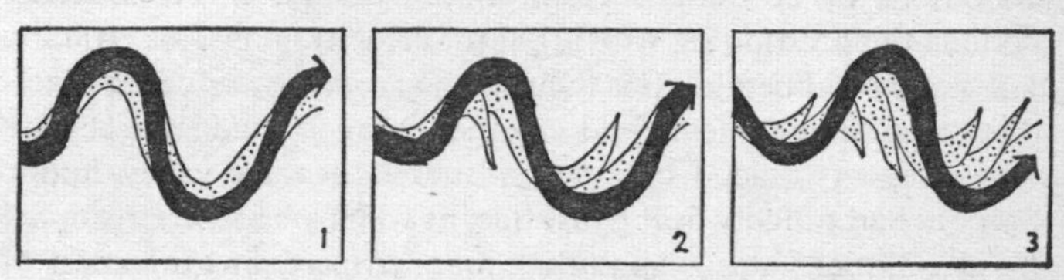


Fig. 48 Sedimentation on a typical flood-plain

flat sheet which is interrupted only by the river-channel and by the remains of cut-offs.

In whichever of these two ways the bulk of the flood-plain alluvium is deposited, it is likely to form no more than a thin skin of superficial material. The solid rock beneath can be neatly planed off, by the shifting meanders, at a depth corresponding to the maximum depth of the pools at bends. If the alluvium does not rest on a flat floor, and if it is far deeper than would be expected, some additional process has been at work. For instance, the valley may have been infilled at some former time. We shall see later, in fact, that many valleys contain quite deep fills of alluvium.

On a large braided channel there will be many braids. These are liable to be destroyed and re-shaped. Destruction takes place at low flows, when the water can break across a braid from one side to the other. Formation takes place at high flows, when many individual braids assume a kite-like or ogee plan.

The very high width/depth ratio of a braided channel means a low level of efficiency. It is obvious that the highest flow velocities, occurring somewhere in midstream and not far below the surface, are also found very little above the channel bed. As a result, shear on the bed is powerful. It is shear that deforms the bed material into braids. The pattern of deformation proves to be random. So also does the pattern in which low-water channels split and reunite. Random behaviour in nature is probably easier to perceive here than in the more complicated situation of the meandering pattern of single channels.

Braiding does not necessarily indicate an excessive total load, despite the great bulk of bed-material that is visible at low stages of flow. Like meandering channels, braided channels correspond to conditions of near-equilibrium among a number of variables. Among these variables, width, depth, velocity, slope, and friction are the most influential. It is their effect *in combination* which determines the three-dimensional channel form. If load were always excessive, all braided channels would infill their valley floors. Some are admittedly doing just this, as when a braided stream fed by glacial meltwater is spreading an ever-increasing thickness of sediment from one valley-wall to the other. But the Rakaia (Plate 38) has performed intermittent cutting. It has left the remains of earlier valley-floors as the tops of terraces, well above existing stream level.

For braids to occur, then, the following are necessary: abundant and preferably coarse bed-material, and a high width/depth ratio. This latter can be regarded as a usual accompaniment of abundant coarse load; but in some cases at least, it may result chiefly from weakness of the banks.

One example of the conversion from meandering to braided has been recorded in detail. The original channel developed a very deep pool on a highly distorted meander. The bottom of the pool penetrated a deposit of ancient, coarse alluvium. The riffle of coarse sediment which formed next downstream split the channel into two. Each arm of the split channel, being reduced in size and therefore in efficiency, increased its slope by cutting. The riffle dried out enough to be colonized by vegetation, and so became fixed. Sufficient coarse material was available to cause further splitting downstream. A meandering reach was in this way converted to a braided reach. Its slope was increased, in compensation for the loss of efficiency.

Incision by braided streams often produces inset terraces. But meanders frequently enlarge themselves during incision, growing sideways rather than shifting down-valley, and becoming ingrown (Fig. 47*b*). Fine examples occur on the lower Seine, where steep undercut banks on one side face gentle slopes on the other. The steep and gentle forms alternate on both sides, all the way along the river. Working backwards in time, we can imagine the process of ingrowth as reversed. It would seem possible to reconstruct a river at higher levels and with less marked curves.

A partial reconstruction can be made. Incision has been intermittent, and terrace deposits record both the altitude and the plan of the valley at former times – that is, at various heights above the present river. Unfortunately the oldest deposits are scattered and fragmentary, for they have been largely destroyed as the valley has been deepened and widened. Thus it is impossible to reconstruct the trace followed by the stream at high levels – in particular, near the general level of the surrounding country. Similar difficulties are encountered in similar situations elsewhere. Consequently it is usually impossible to say whether an incised meandering river had developed bold loops before incision began. There is no doubt that the loops have been enlarged during incision; but a dispute has arisen between those who believe that incised meanders are always inherited from freely shifting meanders on a flood-plain, and those who believe that meanders can originate on a stream which, when incision begins, had a straight trace.

The principle involved in this dispute is an important one in the formal study of earth-sculpture. If the first view is correct – namely, if incised meanders are always inherited from some high-level flood-plain – it must follow that the original loops were developed under the control of a high sea-level, or alternatively that the land has risen. If the second view is correct – if meanders can originate during incision – no movement either of the sea-level or of the crust need *necessarily* have occurred.

The dispute can easily be settled by observation. Plate 37 is a general view of new-formed gullies on a heap of quarry spoil. The slope angle is about 33°. Several of the gullies meander, at least in part: and the marked gully meanders along most of its length. Plate 34 is a close-up view of the same kind of thing, on a heap of spoil where the slope is about 30°. This second example clearly displays the interlocking spurs of ingrown meanders. At the outset, all the gullies in question pursued straight courses downslope. Their meanders have developed during incision.

Similarly for natural rivers. Some reconstructions of former traces at higher than present levels indicate that, when incision began, the channels were approximately straight. We are bound to conclude that ingrown meanders are no necessary criterion of inheritance from some former flood-plain at high level.

The gully examples are especially instructive, since the initial channel slopes were about as steep as one would expect from constant slopes in general. Meanders, then, can form on very steep slopes indeed. Perhaps in some cases we ought to regard meandering as a means of reducing slope. The channel distance on a meandering stream is 1·5 to 2 times as great as the direct downslope distance. Once meanders have formed at all, the conservative nature forced on them by the probability of behaviour could perhaps make them persist indefinitely. Here, however, is yet another loose end in our understanding.

CHAPTER NINE

# *Waves, Bays, and Beaches*

IN close-up, it is quite obvious that wave-attack is highly selective. Joints, bedding-planes, and bodies of weak rock open the way to erosion (Plate 39). The less uniform the exposed rock, the less regular the detailed forms produced by its destruction. In a broader view – wide enough to include the general configuration of the shoreline, and too wide to reveal small details – resistant formations can often be seen projecting in headlands, while belts of weak rock correspond to inlets (Fig. 49*a*). But the forms of shorelines

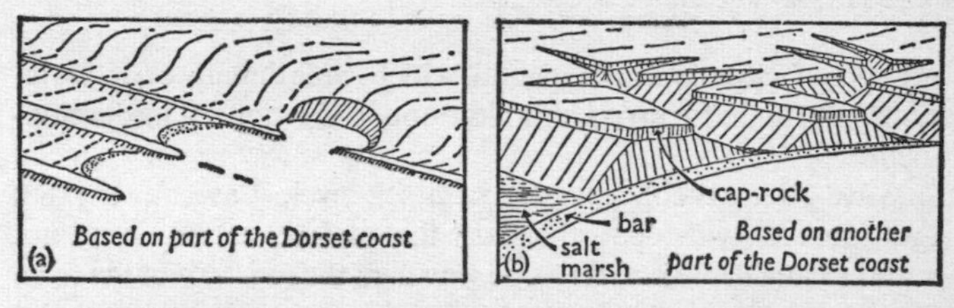


Fig. 49 Varying relationship of shoreline to structure

involve far more than differential erosion of rocks of varying strength. Many inlets are not due to wave-action at all, but are the results of erosion by rivers or by glaciers. Some shores cut indifferently across weak rocks and strong, and across steep hills and deep valleys (Fig. 49*b*). Furthermore, waves can deposit as well as erode. Few pieces of shoreline are wholly devoid of constructional features.

Two reconciliations, therefore, are required. Detailed irregularities of an eroded shore, which delicately reflect differences in rock strength, and arrays of headlands and inlets which correspond to differences in rock-type, must be reconciled with shores which cut regardless across rocks of differing strength and across land-forms of varying shape and unequal bulk. The well-known tendency of breaking waves to destroy the substance of the shore must be reconciled with the undoubted presence, in places, of wave-built features. The first of these two problems is not merely one of scale – of the disappearance of details from a general view – but relates mainly to the stage reached in the development of a given shore. The second problem is one of the dominant behaviour of waves at a particular place. These matters are best approached by way of an

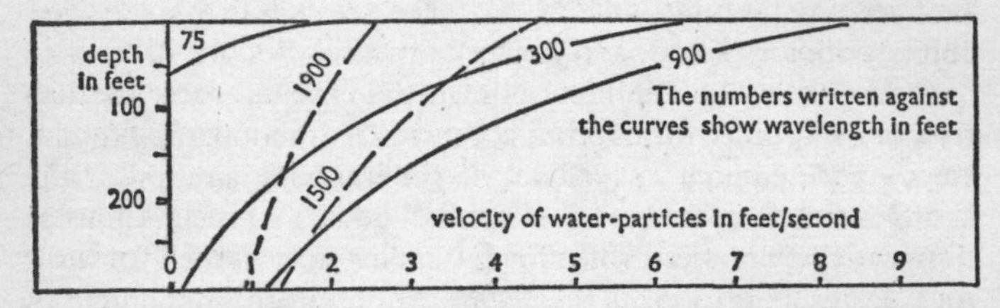


Fig. 50 Speed of water-particles in waves (drawn from data given by Sverdrup, Johnson, and Fleming)

account of wave-action in general, which leads directly on to variations both in the form of a shoreline and in the mode of its development.

Wave-attack is concentrated near sea-level. Wave-motion dies away rapidly with depth beneath the surface, so that it all but vanishes at a depth equal to one wavelength (Fig. 50). In the open sea, where the water is far deeper than one wavelength, the water moves in circular fashion, the size of circles diminishing with depth. But in shallow water, where wave-motion is distorted near the

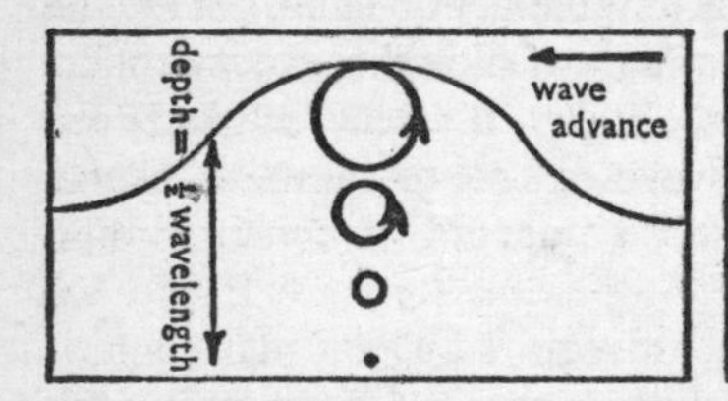


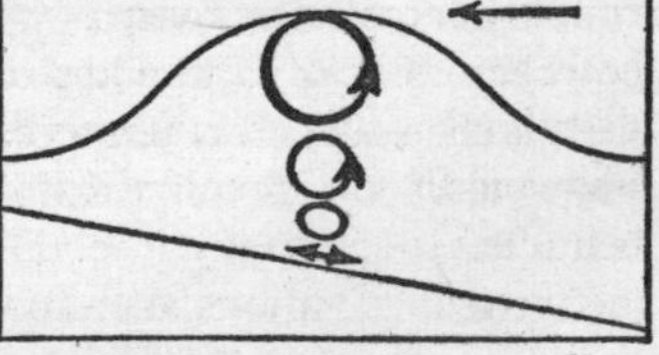

Fig. 51 Motion of water-particles in deep and in shallow water

bottom, circular movement is replaced by movement to and fro: the water swills backward and forward over the sea-bed, even when waves fail to break (Fig. 51).

In order to give some kind of precision to these statements it is useful to quote a few figures. Observed wavelengths of ordinary waves in the open sea do not often exceed 300 feet, and the record is in the neighbourhood of 2500 feet. Very much greater wavelengths are recorded for tide-waves with periods of 24, 12, and 6 hours. Considerable wavelengths are characteristic of the long forerunners of ocean swell, which sometimes exceed 3000 feet between one low crest and the next. But tide-waves and the forerunners of swell have no bearing on the present discussion.

Confining attention to ordinary waves in the open sea, we may note the important distinction between the advance of the wave-form and the actual movement of the water. Waves with their crests 300 feet apart travel at something like 30 miles an hour. This speed is the speed with which the wave-form travels onward, and does not relate to the mass transport of water. In the light of observations of breaking waves on a shoreline, and in view of the march of rollers at sea under a strong wind, it may be difficult to make the distinction required. Oddly enough, it is easy to accept the idea that sound-waves travel through the air, and earthquake-waves through the land, without any net movement of the transmitting medium. Similarly with waves in the open sea – it is the waves, but not the water, which move forward.

In the 30-mile-an-hour waves mentioned above, the surface water travels in its circular orbit at about $4\frac{1}{2}$ miles an hour. At a depth of 60 feet – the equivalent of $\frac{1}{5}$ of the wavelength – three-quarters of this velocity has been lost. At a depth of 300 feet (one wavelength) there is scarcely any circular motion at all. It can be roughly assumed that, on a given piece of shoreline, wave-erosion has little effect at depths greater than half the wavelengths of the dominant waves. It is possible to guess that in many places this wavelength does not exceed 200 feet, so that effective wave-erosion is limited to the zone between high-water mark at the top and 100 feet below low-water mark at the bottom.

If the sea-bottom close inshore shelves upwards towards the land, incoming waves curve over and break. They rush up the beach as swash, beating at high tide against the base of cliffs, and retreating in the broken water of the backwash which flows into the foot of the next advancing breaker. Thus the motion of water on and near the beach is to and fro instead of round and round. In addition, it is violent. Breakers are powerful both in erosion and in deposition.

Waves break because their height increases and their speed is reduced. Increased height makes the wave unstable, and breaking occurs when the velocity of the wave itself falls below the velocity of the water in the crest. The form of incoming waves begins to change rapidly where the depth is about half the wavelength.

Surprisingly little energy is lost by friction on the sea-bed. For slopes between 1 in 20 and 1 in 100 the loss is not more than 10 per cent of the total. Abundant energy remains to hurl the broken water against the land.

Two main erosive processes are to be recognized. One is the undercutting of cliffs, the other is the grinding of beach-platforms. Undercutting is the work of waves which, carrying beach material with them, smack against solid rock. Solid rock on the foreshore, and for some distance below low-water mark, is liable to be ground down by beach material swept to and fro by breaking waves, or rolled backwards and forwards near wave-base (Fig. 51).

Undercutting of cliffs is usually concentrated near high-water mark, where a notch is characteristically found (Plate 39). The cliff is undermined, as if cut by a horizontal bandsaw. Sapping is confined to a depth of a few feet at the most. Within the notch the bare rocks are battered into roundness; at higher levels they display the angular forms produced by rapid weathering and rock-fall, while below the level of the notch they are either concealed by loose beach material or visibly truncated by the gentle seaward slope of the beach-platform.

Little can actually be said about the precise relationship of the height of the notch to the level of the sea. The question has much practical importance, for old notches can be used in determining former sea-levels (p. 119). Even if it is assumed that the notch marks high-water level, uncertainties are still to be met. A very short inspection of an actual piece of jagged shore shows that the present notch slopes quite sharply upward along the edges of minor inlets. Notches are being cut on some shores at the present day well below the level of high water; Plate 40 is a photograph of a notch some 20 feet below high-water level and 10 feet below mean sea-level. On this part of the shore of Guernsey the rocks are traversed by widely spaced but well-defined joints. Where a block has been destroyed, the large angular hollow formed by its removal contains coarse cobbles. These, moved by waves at low-water stages, are cutting notches in the bases of remaining blocks. Observations of this kind show how difficult it is to relate present wave-action precisely to present sea-level, to say nothing of reconstructing the sea-levels of the past.

Waves breaking freely against the land carve distinctive erosional features. Caves are opened along the lines of master joints or in the bodies of weak dykes. Although little or no direct observation has been carried out, it is thought that caves can be enlarged beyond the limits reached by the waves which drive into them. Air, compressed before the invading water, is forced under pressure into the confining rock. When the wave retreats and the mouth of the cave is suddenly freed, the pressure is released. Air compressed in fissures expands under the reduced pressure, flaking off pieces of rock. This process is comparable to ice-wedging (p. 192). It is thought possibly to be responsible for blow-holes – holes which connect the interior of caves with the cliff-top.

Even if no caves are opened, recesses are often cut, with the cliffs projecting in buttresses between them. A long, narrow buttress, undercut and partly collapsing in the middle, becomes an arch (Plate 41). Further collapse, or the simple detachment of a buttress from the main cliff, produces a stack (Plate 42). But in the general view and also in the long view, caves, arches, buttresses, and stacks are no more than details in the form of a receding line of cliffs.

Being quarried away at the base, cliffs tend to overhang. But some rocks are far too weak to support vertical slopes, to say nothing of projections, and the actual profiles of receding cliffs vary a great deal (Fig. 52*a*). Vertical cliffs in full retreat rise in sheer walls, as on the two sides of the Straights of Dover where flat-lying Chalk presents abrupt faces to the sea. Sites of this description well illustrate the independence of shorelines on the form of the land, for the ends of hills are smoothly truncated at the shore. In north-west Ireland half a mountain has gone (Plate 43), the present summit of Slieve League lying at the very top of cliffs nearly 2000 feet high.

At many places, however, it is only the lower part of the cliff which is directly affected by undermining and collapse – the upper part is shaped by some other agency. It is quite possible for the upper part to be weathered back as fast as the base is undercut, so that the two elements – the vertical base and the sloping upper portion – recede at a single rate. In that event, the upper facet is a constant slope of the kind discussed on p. 69. But on either side of the English Channel, where dog-leg cliff-profiles are very common, some allowance must probably be made for weathering and erosion in cold climates (p. 189), as well as for complications resulting from movements of sea-level.

As cliffs recede, so is the beach-platform extended. But, because sea-level has been unstable for at least a million years, there is no

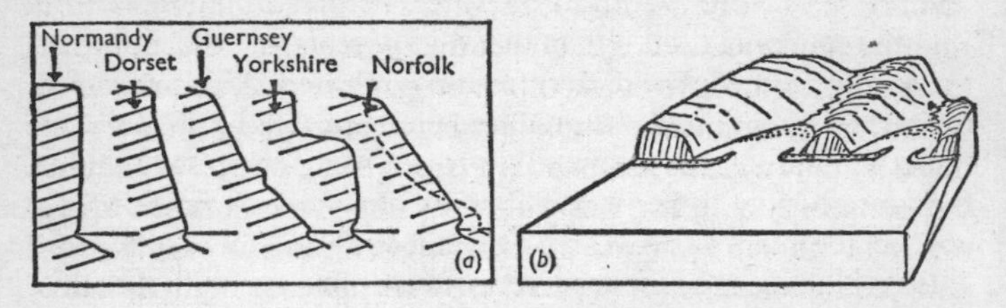


Fig. 52 (*a*) Some cliff-profiles
(*b*) Spits and headlands

means of telling whether, if that level remained constant, the sea could encroach on the land to very great distances. The widest platforms ascribed to abrasion lie at modest heights above present sea-level, and do not exceed a few miles in width.

In the central Pacific and in the southern oceans generally, numerous platforms occur a few feet above present sea-level, on shores where the tidal range is slight. In the main, these platforms are narrow. Many are subjected to washing-over by storm waves, which can reach the feet of some of the cliffs behind. Accordingly, some writers class the features as storm-wave platforms; but others, impressed by remarkable horizontality in some places, emphasize the effect of weathering – especially, weathering in and on the margins of rock-pools. The apparently obvious view, that these platforms relate to a former stand of the sea, a few metres above the present mark, does not command much respect among the investigators mainly concerned.

When the shoreline is considered in plan instead of in profile, its features arrange themselves in some kind of order. The kinds of features which occur depend partly on the form of the shoreline under attack and partly on the progress which erosion has already made. For example, if the coastland is deeply cut by wide valleys and diversified by substantial hills, the shoreline may well be indented by ample bays separated by bold headlands. Debris cut away from the headlands and brought down by rivers is pounded by the breaking waves into the sand and shingle of beaches. Confined at first to narrow strips below the cliffs, and to crescents at the heads of bays, the beaches grow. Spits projecting from headlands tend to grow right across the mouths of bays (Fig. 52*b*), until in time the inlets are sealed off from the open water. What happens next depends on the balance between cliff-retreat and the accumulation of sediment. If time permits, the enclosed lagoons are filled by blown sand, peat, and the deposits of rivers. But the cliffs are constantly receding, allowing the shore to move inland across the inlets as well as across the headlands. Eventually the lagoons and the lagoon deposits will be destroyed, and a continuous line of cliffs will mark the limit of the land.

Extension of spits is one of the results of longshore drift, the zig-zag movement of beach material. Oblique waves wash this material obliquely up-beach, and drag it directly down-beach, in alternating swash and backwash. Groynes can be used to arrest drift and to widen the beach (Plate 45). Where oblique waves approach from two directions, spits can grow out from both sides of an inlet. Most longshore transport, however, is effected in the surf zone. Waves stir up sand, which is then shifted along the offshore belt by the longshore current.

If the sea-bed slopes very gently away from the land, large waves are forced to break offshore. Plunging on to the sea-bed in shallow water, they throw up sand and shingle in the form of offshore bars (Plate 44). Although the seaward face of a bar is combed down and destroyed, the inner face is renewed by the washing-over of debris. Thus the bar retains its identity, but is forced towards the land.

A huge series of bars fringes most of the North American coast from Norfolk, Virginia, to Panama – a distance of 4500 miles.

Shallow, branching inlets on the Atlantic coast of the U.S.A., protected from the open sea, are being filled by mud and peat beneath the marsh-plants of the Dismal Swamps. The Dutch Polders and the English Fenland result from complete infilling. However, lagoons of this kind, like the lagoons previously described as formed from bays, are open to destruction as the bar is driven farther and farther in. When it rests everywhere against the land it forms a line of beach beneath low cliffs.

Theoretically, an offshore bar or a cliff-backed beach can run straight for very long distances. In actuality, individual stretches of bar, beach, and cliff are usually curved. Their form is controlled not by waves in general, but by certain waves – dominant waves – in particular. The height, frequency, and wavelength of dominant waves are strongly affected by the length of fetch – i.e. the distance of sea traversed by the winds which raise the waves. Lines drawn, radius-like, from the curve of a particular beach often point across a particularly wide extent of sea (Fig. 53).

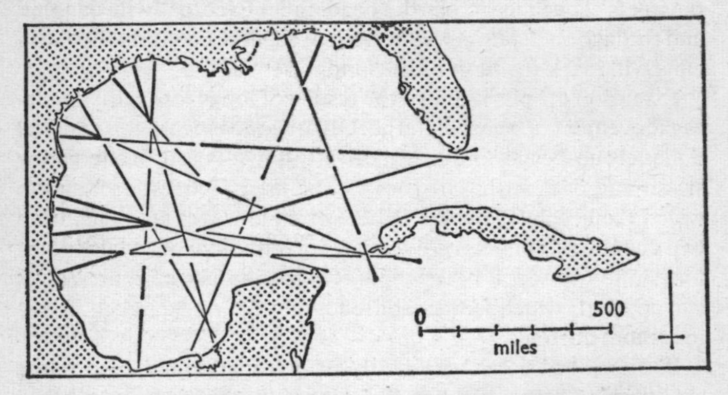


Fig. 53 Orientation of beaches and directions of greatest fetch, Gulf of Mexico

In the accompanying graph (Fig. 54) wave-height is related to length of fetch. It is also related to duration – i.e. to the duration of the wind which raises the waves. The graph is drawn for a wind-speed of roughly 30 miles an hour. A wind blowing at this speed over a fetch of 100 miles could raise waves 12 feet high – provided that it blew for 10 hours; but if it blew for only 5 hours, the waves would only be 8½ feet high. A 30-mile-an-hour wind blowing for two days across 900 miles of sea raises waves 17 feet high.

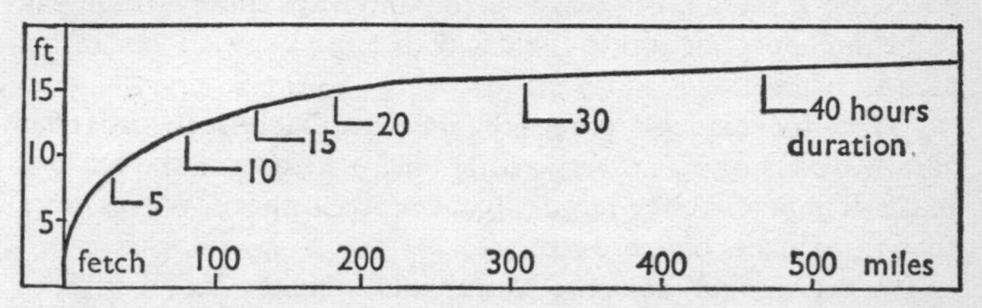


Fig. 54 Maximum height of waves raised by 30-m.p.h. wind (adapted from Sverdrup, Johnson, and Fleming)

Any one piece of shore presents a unique problem in the combined influences of wind-speed, wind-frequency, wind-direction, wind-duration, frequency, wavelength, wave-frequency, and wave-height. Allowance also has to be made for refraction of waves round headlands. Furthermore, the combination can change from season to season. In middle latitudes, waves of greatest length, height, and energy are characteristically those generated by the winds of cyclonic storms during the winter season. Such waves are typically destructive. They tear down beaches, steepen beach profiles, reduce beach widths, and move beach material offshore. Summer waves, shorter in wavelength and less in height and in steepness, deliver less energy. They move the beach material back toward the land, building the beaches up again, widening them, and reducing their seaward slopes. Hence the usual contrast between winter and summer beach profiles, and in the appearance of the beaches in general. The seasonal changes are complicated by changes within seasons. Offshore winds, for instance, tend to flatten waves and so to reduce their energy and to make them constructive.

The most violent waves of all are the tsunamis generated by earthquakes, and those associated with tropical cyclones. But tsunamis are infrequent, tropical cyclones are restricted in their areas of impact. Widespread and highly destructive waves include powerful swells generated in midlatitudes, especially the southern ocean swell that beats on shores in South America, Africa, and

Australia. With wave periods of 12 to 15 seconds, the southern ocean swell rises in breakers up to 20 feet high along the south-eastern edge of Australia. Fetch over the southern ocean is effectively limitless. But wave height, and wave energy, decline in logarithmic proportion to the decay distance between generating area and shore.

At the other extreme, part of the west Florida shore records near zero wave energy. The very gentle offshore slope of the sea bed makes it impossible for large waves to come inshore. Waves more than three inches high are rare.

The alternation between winter and summer beach profiles represents an alternation between two sets of steady-state conditions. Disequilibrium is represented by a progressive tendency for a beach either to build out or to cut back. Such a tendency can obviously result from a change in wave energy, itself related to changes in the frequency and location of storms. On the whole, shoreline erosion attracts more notice than shoreline construction, on account of the property and amenity damage involved.

A remarkable assemblage of constructional features is observed on the eastern side of Romney Marsh, in Kent, where a broad projection of low ground terminates in Dungeness. The point of Dungeness is backed by gently curving ridges of shingle, which are being progressively colonized by plants. Further inland, beneath the main expanse of the Marsh, older ridges are concealed by a spread of alluvium. At the present time the Marsh is growing towards the east but being cut away on the south (Fig. 55*a*). The

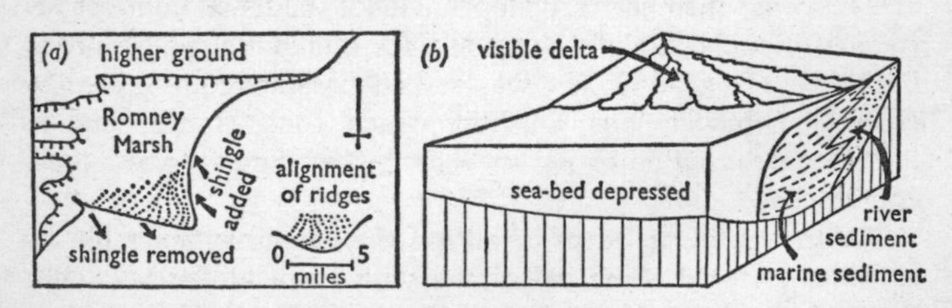


Fig. 55 (*a*) Shingle bars on Romney Marsh
(*b*) Form of large delta

southern ends of old ridges are being removed about as fast as new ridges are built on the eastern side. The rate of growth towards the east is roughly equal to two ridges a year; it may be wondered if the two new ridges are thrown up at equinoctial spring tides.

Since a large fraction of the material which lies on existing beaches has been brought down by rivers, rivers make a great indirect contribution to constructional activity on the shoreline. They also contribute directly, by filling inlets with sediment and by building deltas (compare Plates 46, 47).

The type delta is that of the Nile, which was named for its similarity in plan to the Greek capital letter *Δ*. Other deltas – including those of the Rhône and the Po – are also located on the shores of the Mediterranean Sea. Now because of the sizes and shapes of its constituent basins, the Mediterranean develops very small tides – the natural wave-periods of the basins do not accord with the 12½-hour period of tides, or with harmonics of it. In consequence, the belief has arisen that deltas form only in seas where tides are small or absent. That belief is quite unfounded.

The great delta of the Colorado (Plate 47) lies at the head of the Gulf of California. In direct contrast with the Mediterranean basins, this Gulf is so shaped as to amplify the tides. Tidal range and tidal currents are known to reach maximum strength at the very mouth of the river. Ranges in the river itself vary from 12 to 15 feet. Tidal bores produce wave-forms 6 feet high. The essential facts were recorded by Europeans more than 400 years ago. Francisco de Ulloa, one of Cortés's men, observed in 1539 that the violence of the tides caused the sea to run with so great a rage into the land that it was a thing to be marvelled at. He also saw that the tide turned back with a like fury at the ebb. In these conditions tidal scour is bound to be severe. Nevertheless, the Colorado has formed a delta well over 3000 square miles in extent.

It cannot be disputed that the range of 30 feet is high. The comparable range – i.e. the range at high spring tides – is 27 feet at Liverpool and only 21½ feet at London Bridge. Both of these places are noted for their great tidal range. But the Colorado may seem an unfair case, since it carries a notoriously heavy load of sediment, much of which lies within the sand-grade and is likely to settle quickly at the river-mouth.

Other deltas, however, also record high tidal ranges. At the entrance to the Amazon Delta the range is nearly 14 feet at spring tides, rising upstream to 16 feet. Although these values are below those quoted for London and Liverpool, they still exceed the 13 feet at Southampton. Delta-building at the mouth of the Amazon is, perhaps, especially instructive, as the gradient of the lower river is very slight, and as a high proportion of the total load of rock-waste is brought down in solution.

A long list of deltas and tidal ranges could be given, but for the sake of brevity no more than three additional instances will be cited. On the delta of the Fraser, in British Columbia, the diurnal range of tides exceeds 10 feet. At the mouth of the Ganges the range at spring tides is 15 feet, and at the mouth of the Irrawaddy it is 18 feet. Enough has been said to warrant two conclusions: deltas do form on the shores of highly tidal seas, and writers who state the contrary have never checked their statements against tide-tables.

All that is necessary for delta-building is an excess of deposition over removal. Except in very special conditions of scour, deltas are likely to form at the mouths of all rivers. Coastal deltas would be far larger and far more numerous than they actually are, but for the operation of three processes. First, wave-action tends to smooth away projections, including the projections of deltas. Secondly, under the weight of large deltas the crust tends to sag, so that the *visible* extent of the features is reduced. Thirdly, the world-wide rise in sea-level resulting from the decay of ice-caps (p. 158) has partly submerged many deltas and may have entirely concealed others; here again, their visible extent has been diminished.

Waves can destroy the longest projecting arms of a delta, and longshore drift can carry loose material away from the place where it is first deposited. But such destruction and removal are at least partly compensated by the infilling of bays between the branching arms. Various patterns of spits and lagoons result from the combined action of deltaic deposition and shoreline processes, and these patterns can, if desired, be used to classify deltas according to differences of plan.

Large deltas depress the earth's crust under their weight of sediment. The best-suited delta in this respect is that of the Mississippi, which is known to have been subsiding intermittently for many millions of years. Each downward movement brings marine conditions inshore, whereas the intervening episodes of stability permit the delta to grow seaward. The net result is an interleaving of marine, deltaic, and river sediment (Fig. 55*b*). Subsidence is not confined to the piece of crust immediately beneath the delta. Adjacent portions of the shoreline are bent down and drowned. Submergence on the flanks of a great delta is well illustrated by Zeeland and the Zuider Zee area of the Netherlands, where land exists only because it has been reclaimed from the sea.

The general rise in sea-level associated with the decay of ice-sheets is independent of any local subsidence. It has compelled delta-building to begin afresh with reference to a new datum. Were it not so, coastal deltas might well have been so extensive and so common that the myth connecting them with tidelessness might never have arisen.

## CHAPTER TEN

# *Ups and Downs of the Sea*

THE discussion of planation in Chapter Six was limited to erosion by rain and rivers. Chapter Seven noted, in the context of long-profiles, that the relative level of land and sea is by no means necessarily stable. We come now to the question of marine planation. Such planation is possible, but its possible extent remains disputed. Its advocates – probably a minority, although a vocal one – rely on wave action to produce broad and gently-sloping platforms. The erosional character of the platforms is certain enough: they cut regardless across rocks and structures.

Erosional platforms slope so gently that they can easily be taken for old shoreline features, whatever their origin may actually be. Seaward gradients on the benches now being cut on existing shores typically range from 10 feet per mile to 80 feet per mile. A gradient of 10 feet/mile corresponds to a difference in height of 250 feet over a distance of 25 miles, and that of 80 feet/mile to a difference of 2000 feet over a similar distance. The arithmetic means of these extreme gradients, i.e. 45 feet/mile, would produce a seaward fall of 1125 feet if continued for 25 miles. Considerable ranges in height, therefore, do not in themselves prevent a given erosional platform from being classed as marine. The gradients mentioned are far greater than the gradients typical of large rivers on low-lying plains, and are certainly not too low to be the gradients of the general surfaces of such plains. Slope alone, then, is not enough to separate terrestrial from marine surfaces of planation.

Where high-standing erosional platforms have been much dissected, it is often difficult to see how they are related to one another. But where they are reasonably intact, they usually end in fairly abrupt ascents to the next higher levels. If a given platform was originally wave-cut, its steep inner edge represents a former line of cliffs. In that event, its present height is due either to a rise of the land, to a fall in the level of the sea, or to both combined. Even if the platform is part of an old landscape, which was reduced to a very subdued state in the final stages of a previous erosion-cycle, its present height may be explicable only by some change in the relative level of land and sea.

The problem of erosion-platforms is by no means local or special. On the contrary, it is world-wide and general. As soon as one realizes that highland areas do not usually consist of a random assemblage of sharp peaks, but are frequently typified by broad gently sloping surfaces at high levels, it becomes obvious that platforms demand explanation. Even where the high ground is deeply cut by numerous valleys, erosional flats of varying size are likely to be encountered. There can be no doubt that large-scale and intermittent movements have taken place, whether movements of the land or movements in the level of the sea.

Now if a shoreline platform is left stranded by a fall in sea-level, without any local warping of the earth's crust, the old shoreline remains horizontal. In vertical plan, it is parallel to the present shore (Fig. 56a). Horizontal series of old beach-platforms and of

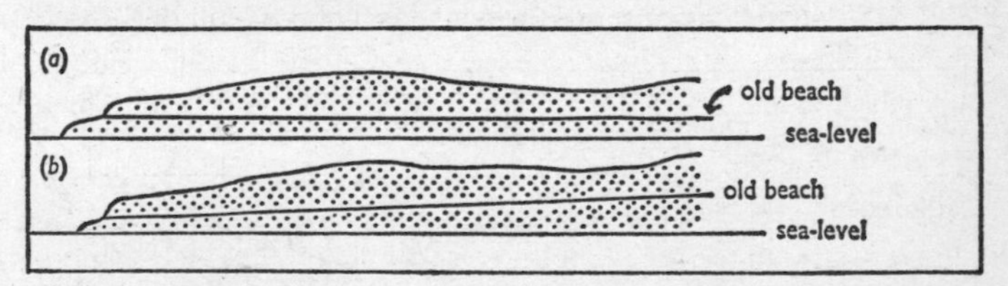


**Fig. 56** Raised beach (*a*) parallel (*b*) not parallel to present beach

old beach-deposits were identified in the Mediterranean more than 60 years ago. In some localities the features in question are remarkably well defined. They lie at heights of 300, 180, 100, 60, and 25 feet above the present shore. But although they prove general changes in the level of the Mediterranean, the suggestion that similar changes have affected all the oceans of the world was strongly resisted for a long time. Not until benches at similar heights had been traced along the Atlantic and Channel coasts of France were the full implications of the old Mediterranean shores recognized. By that time it had been shown that, throughout the relevant period, the Mediterranean had been connected with the Atlantic. The level of the Mediterranean was necessarily controlled by the level of the Atlantic outside the Straits of Gibraltar; it could no longer be urged that the recorded changes in level had been confined to an inland sea.

Nobody familiar with the terraces of the Thames can fail to observe the significance of the heights noted for the Mediterranean platforms. Three main terraces have long been identified in the lower Thames valley, at heights of 100, 60, and 25 feet above the present alluvium. Since these three terraces are the remains of old flood-plains, it follows that they were deposited with reference to former sea-levels, at heights of 100, 60, and 25 feet above the present mark. Although recent detailed work has increased the number of terraces identified, it does not contradict the agreement between the Mediterranean and Thames-side sequences – particularly since an additional terrace has been located at about 180 feet above the present Thames.

On the Atlantic coastland of the U.S.A. two clearly marked series of old shoreline features stretch from New Jersey to Florida (Fig. 57), with extensions into the coastland of the Gulf of Mexico.

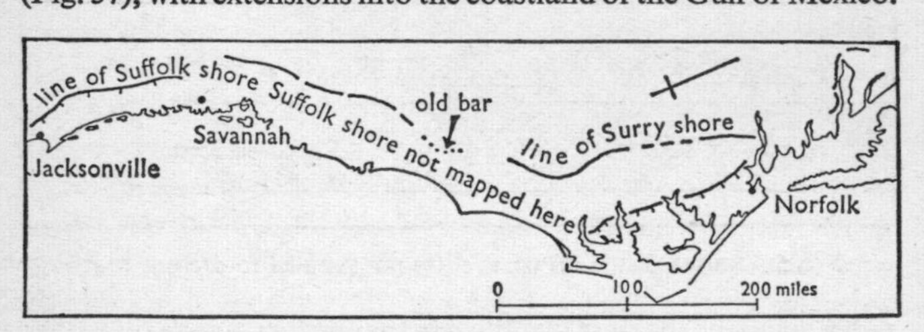


**Fig. 57** The Surry (100-ft) and Suffolk (25-ft) shores (adapted from Flint)

The lower of the two, the Suffolk Scarp, consists chiefly of an old cliff-line about 25 feet above present sea-level. In places, the cliffs are 60 feet in height – i.e. their tops reach 85 feet above sea-level, their feet remaining at 25 feet. Elsewhere – notably in South Carolina – cliffs are replaced by bars of beach-material comparable to the great bars lining the present shore. In all, the Suffolk shore has been traced over a length of 800 miles.

Behind the Suffolk shore, lying 100 feet above present sea-level, comes the Surry Scarp. This second shoreline has been traced for nearly 400 miles. Although its features are somewhat degraded, they are recognizably coastal.

Each of these two North American shorelines agrees in height with one of the Mediterranean levels. A 25-foot beach is also known from Bermuda. A wide, gently sloping bench at similar height is recorded from the Pacific coast of Kamchatka, in eastern Asia, where a shore 100 feet above existing sea-level is also present. In Australia and Tasmania the 100-foot shoreline is widely reported.

Although it cannot be claimed that old shorelines at specified heights are known throughout the world, the scattered evidence so far collected is at least widely distributed. Such general agreement as can be claimed relates, however, mainly to old shoreline features at 430 feet or less above sea-level. For reasons which will appear presently, higher platforms still provoke keen discussion.

One obvious possible cause of general changes in sea-level consists in the waxing and waning of great continental glaciers during the Ice Age. At the present time, glaciers and polar ice combined hold in storage about 2 per cent of the world's water. The oceans hold more than 97 per cent. When glaciers grow, it is only from the ocean that water can be taken, in order to form ice. It is estimated that, at glacial maximum, some 10 million cubic miles (40 million $km^3$) of water are locked up as 11 or 12 million cubic miles (45 million $km^3$) of ice. The abstraction of water from the oceans causes sea-level to fall, perhaps as much as 500 ft (150m) below its non-glacial mark. The reckoning includes an allowance for the sagging of vast land areas under their heavy glacial loads.

Glaciers today are only about one-quarter as bulky as they were at former glacial maxima. But they still retain enough water to bring sea-level about 100 ft (30m) above its existing mark, if they were all to melt away.

Since there have been at least four major peaks of glaciation, sea-level has been depressed and subsequently elevated at least four times. The last fall was partly responsible – and perhaps mainly responsible – for making rivers cut down towards the lowered shore. Mouths of valleys were rapidly excavated. When the ice melted off, the sea-level rose, the new cuts were invaded by salt water. Drowned valley-mouths are common throughout the world. Their form depends on the texture of the drowned landscape. Where this was at all rugged, or at least contained deep-narrow valleys, the coastal inlets produced by drowning are narrow, deep, and steep-sided. Such inlets are common in the south of Ireland, south Wales, southwest England, northwest France, and northwest Spain (Plates 33, 48).

Drowned shorelines are much harder to identify than are drowned land-surfaces, simply because they lie wholly under water. One submerged shore, identified some time ago now, lies off the eastern coast of the U.S.A., at a depth of about 300 feet (100m). Others at lesser depths have been studied, and continue to be studied, for instance round Hawaii and in the area of the Bahamas. Scuba diving and the use of small submersibles provides far more abundant information than was formerly available.

Certain peat-beds, and groups of tree-stumps in the position of growth, are widely known from coastal districts at levels well below that of the present sea. Naturally enough, many of the relevant sites are located in drowned valleys, where borings for bridge foundations and excavations for harbour works bring the evidence to light. Peat, containing tree-stumps still rooted in it, has been dredged from the floor of the North Sea at depths as great as 170 feet (55m). Remains of inter-tidal shellfish come from even greater depths on the Long Forties Bank off eastern Scotland.

Interpretation of the evidence can be complicated where the crust, relieved of a load of ice, is still rising. But even there, the rise in sea-level may for a time be faster than the rise of the land. Evidence of submergence from areas where the crust has long been stable conclusively prove a rise in sea-level. Most is known about the latest rise. This began slowly, about 15,000 years ago, then gathered speed before slowing again and finally ceasing about 5000 years ago, when something very close to the existing sea-level had been attained.

Now if world-wide changes in sea-level are due solely to the growth and decay of land-ice, each rise should compensate fully for the preceding fall. The fact that the world's glaciers are still quite bulky does not affect the argument. Let it be admitted that existing sea-level is below the level when the Ice Age began. The fact remains that the major extensions and retractions of the land-ice appear to have been quite similar in area, for the last three or four fluctuations at least. Existing sea-level ought therefore to be at about the mark recorded prior to the last major extension of the glaciers. In practice, it is found to be lower. And restoration of sea-level to a non-glacial mark, by means of the melting of the ice which remains, would still leave many platforms, some of them carrying old beaches, stranded out of reach of the waves. Something has happened, quite in addition to the glacially-controlled fluctuations.

When former high sea-levels are plotted against time, they show a falling trend during the last half million years or so. The rise of level resulting from deglaciation failed to restore the sea to its immediately previous high mark. It is not very easy to define a firm trend, except that of a progressive fall of level, on which glacial/interglacial fluctuations are superimposed. But against this, it is not easy to imagine how older and older sea-levels could go ever higher and higher. An attractive interpretation is that of Emiliani, who suggests that non-glacial sea-level may now be close to the trough of a cycle (Fig. 58). On his view, the cycle crested somewhere near 500,000 years ago, and has now perhaps just passed its trough.

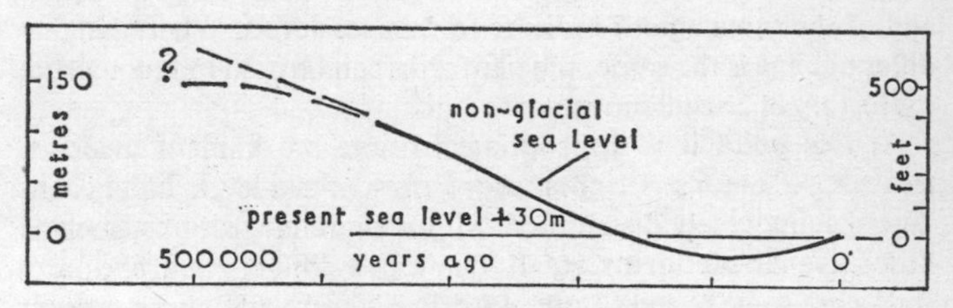


Fig. 58 Possible cyclic fall of sea-level during the last 500,000 years (adapted from Emiliani)

If this interpretation is correct – or even if we infer a fairly steady fall during the same interval of time – we need to consider a change in the capacity of the ocean basins. Beyond this point begins guesswork, with guesses inevitably related to the operation of processes at work in the earth's interior (Ch. 11).

One general difficulty is that of dating. To assume that equal height means equal age is to assume the very thing that requires testing. Imagine two platforms, say at 200m above the present sea, but on opposite sides of an ocean, or even on opposite sides of the world. They can only be related to a former stand of the sea at +200m, if the crust in which they are cut can be shown to be stable, and if they can be proved to be of the same age, by independent means. In reality, it is proof of identical age which would give the best evidence of crustal stability.

Dating by fossils, if shoreline or shallow-water fossils occur, can only refer to the localities themselves. At that, it is likely to be far too imprecise for the purpose in hand. Dating by radiometric methods offers the most likely – and perhaps the only – prospects. But while radiocarbon dating provides massive information on the last 36,000 years, and some information on the last 72,000, its practicability ends at this latter mark. Other radiometric methods deal with the interval from about 5 million years back to the origin of the earth, at about 4500 million years ago. There are indeed techniques applicable to the interval between 5 million and 50,000 years, but these are fairly restricted in use. They work for instance on the stalagmites and stalactites of limestone caverns.

If precise dating could be effected, its advantages would be several. It would enable any sets of platforms at the same height and of the same age to indicate former sea-levels. Where heights differ but age is the same, the platforms can be used to reconstruct the history of crustal movement.

At this point it is appropriate to note the kind of evidence treated. At modest heights above present sea-level, old beach-deposits are widely distributed. Minor shoreline features, such as little wave-cut platforms, small abandoned cliffs, caves, arches, and stacks are to be found (Plates 49, 50). They are, of course, present on shores where the land has risen as well as on those where the only movement has been a lowering of sea-level. Behind the former beaches, terraces run up the river-valleys. In especially favourable conditions it is possible to relate a given terrace to a given beach. The profiles of some rivers are broken at heights which agree with the heights of old shoreline features (p. 88). Although work on a particular area may well prove difficult and complex, general changes of sea-level are abundantly indicated by both erosional and by depositional features at heights up to 200 feet or 300 feet – that is, within the range of shoreline development during the Ice Age.

Correlation becomes more difficult with increasing height. Old beach-deposits, and the river-terraces which represent old flood-plains, survive best at low levels. At greater altitudes, having been longer subject to erosion, they are either much dissected or non-existent. In the absence of superficial deposits, and of the fossils which they contain, it is always difficult and often quite impossible to set a firm date to a bench or an old shoreline. Height remains the only guide to relative age. Height cannot validly be used in correlation unless, for a given platform, it is sensibly constant. Constant height cannot be expected unless that platform in question is an old coastal feature, with its inner edge representing the old shore. At this point we are thrown back on interpretation.

Above the range of old beach-deposits and minor erosional forms, especially at high levels where platforms are much dissected, more than one interpretation can be placed on a single body of observations. Some workers, used to the fine distinctions possible in coastal districts, tend to subdivide the platform-sequence minutely. Others, taking a broad view, are prepared to believe in great ranges of height, even in the absence of crustal warping.

In the early nineteen-fifties, Balchin identified, on the flanks of Exmoor, old shorelines at 1225, 925, 825, 675, 425, and 280 feet above sea-level, whereas Brown has reached the conclusion that the interior of Wales consists wholly of the remnants of former land-surfaces. Below the high residuals ranging in height from 2100 to 3500 feet, Brown distinguishes the following series of platforms, in descending order (Fig. 59); 1700–2000 feet; 1200–1600 feet; 800–1000 feet. Below the last comes an old shoreline at 650–700 feet, and below that again the 430-foot bench.

Both interpretations imply, and depend on, the assumption that the two areas have been crustally stable – say, during the last 50 to 75 million years. There is one possible exception: a curved envelope which can be fitted to the highest Welsh summits can be read to indicate broad warping. But the plateau remnants between 1700 and 2000 feet above existing sea-level are interpreted as the surviving portions of a very subdued erosional surface, which can scarcely fail to have stood very little above the sea-level of its time. On this view, we must postulate for Wales (and also for other stable areas) a former sea-level of, say, 1600 feet above the

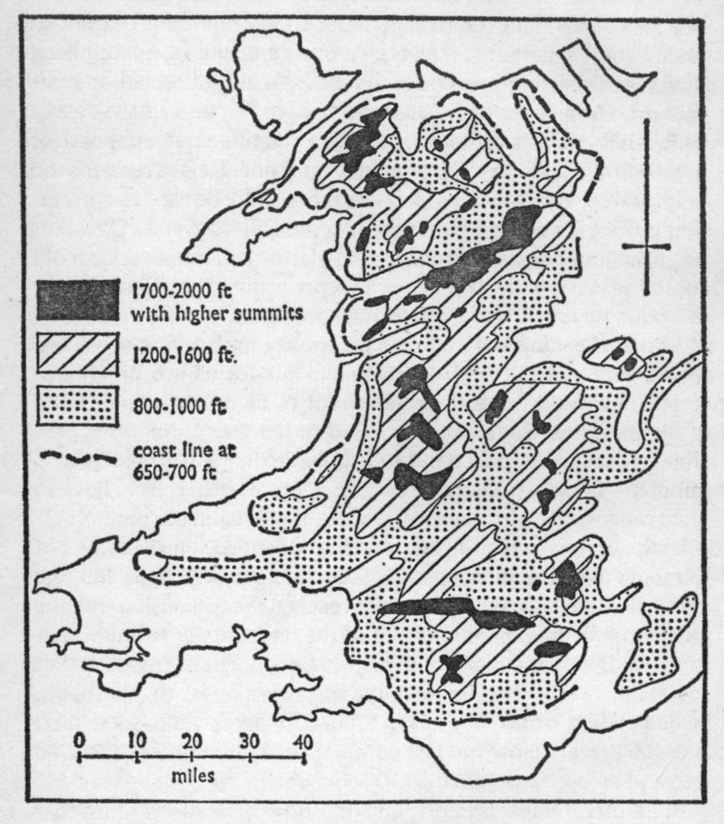


Fig. 59 Erosional platforms of Wales (simplified, after Brown)

present. Because all signs of any shoreline at such a level have long been lost to erosion, neither its location nor its plan can now be reconstructed.

Doubts have been cast on the required stability for the Exmoor area. Some of the risers which ascend from one platform to another occur on lines of faulting. They may thus be structural.

In view of what has been said already, there is no need to examine at length the presence, in southeast England, of platforms at 200, 430, and 600 feet above sea-level, but a few comments on their form and relations may usefully be made. The so-called 200-foot platform is probably to be correlated with the 180-foot beach of the Mediterranean. It is represented in part by erosional flats cut into the flanks of the Chalk downs, in part by high terraces which are recorded both in the Thames valley and in the Hampshire Basin. Detailed work on the South Downs, in the Hampshire Basin, and in the Weymouth area separates a series of features at 290 feet from another at about 325 feet. Assuming that one of these corresponds to the highest of the Mediterranean beaches, more work is needed to show which should be selected. Less well-developed or less widespread flats are present at higher levels, including that of 430 feet, which is widely recorded along the Channel coast. In the London Basin it appears mainly as gravel-spreads deposited by an early Thames. The 600-foot shoreline has been traced all the way round the Hampshire and London Basins, in the form of a platform cut into the encircling Chalk.

At Headley Heath, near Dorking, some of the old shingle remains on the 600-foot beach. But, in general, such deposits are rare. Even where they do survive, they give little help in dating. The fossils obtained from them are poorly preserved and incapable of fixing the age of the deposits within the fine limits required. All that can be said is that the material dates from the Pliocene – the division of geological time immediately preceding the Ice Age (Pleistocene). So much could, perhaps, be inferred on general grounds.

At the 800-foot level and a little above, the London and Hampshire Basins are surrounded by subdued Chalk crests, on which rests a mantle of clay-with-flints. This material is the product of prolonged weathering. It is composed of the insoluble flints, sand-grains, and clay particles which occur as impurities in the Chalk. As the Chalk was dissolved and carried away in solution, the impurities remained behind. Since unweathered Chalk contains about 98 per cent of calcium carbonate and only 2 per cent of other matter, a thickness of 2 feet of clay-with-flints corresponds to the solution of 100 feet of Chalk. Far greater thicknesses than 2 feet of clay-with-flints are known in some places. It can therefore be inferred that the Chalk summits have been severely affected by weathering which continued for a very long time indeed.

In order to account for prolonged weathering, one must infer that the Chalkland was long exposed to the air and to the erosional forces operating on land-areas. Since the summits near the 800-foot mark vary little in height among themselves, they are taken for the fragments of a former landscape which used to be much more nearly intact and much more extensive than it is today. Its subdued relief is ascribed to erosion under the control of a stable sea-level – a sea-level which remained stable long enough for the land of southeast England to be reduced to a very subdued condition.

The stronger rocks of the west resisted erosion at the 800-foot level more successfully than did the rather weak sediments of the southeast. They have also resisted later erosion well enough to preserve quite extensive remnants of the 800-foot shoreline, of the coastal belt adjacent to it, or of both. The Welsh platform at 800–1000 feet is presumably to be correlated with the Chalk summits of the southeast. The shoreline at 825 feet on the flanks of Exmoor seems likely to belong to the same part of the sequence, even though it seems a little high. Farther afield, old coastal platforms at about 800 feet above the present level have been located in the west of Ireland. In Donegal, their remains are adjoined on the inland side by gently undulating country which seems to belong to a land-area subdued by heavy erosion.

Without going beyond the limits of the British Isles, we can see that reasonably convincing evidence exists in quite widely scattered localities for a sea-level about 800 feet higher than the present, and for extensive and long-continued erosion of the land which produced subdued platforms at slightly greater heights. The 800-foot platform is crucial to the understanding of the southeast, for very few points rise above it. If it can be dated, it sets an earliest possible limit to the erosion effected in southeast England since the crust in that region was last deformed – it cuts across gently tilted rocks itself remaining horizontal, and manifestly post-dates the last extensive earth-movements. If it can be dated, again, it provides a

30. Neretve Valley, Yugoslavia

31. Hillside slopes on Carboniferous Limestone, Denbighshire

32. Tree undercut by meander, Epping Forest

33. Meanders on the Towy, South Wales

34. Meandering gully on spoil-heap

35. Banks and backswamps on artificial stream

36. The Rift valley of the Jordan

37. Gullies on a spoil-heap

38. Braided channel of the Rakaia, N.Z.

39. Wave-cut notch, Dorset coast

40. Notch near low-tide-mark, Guernsey

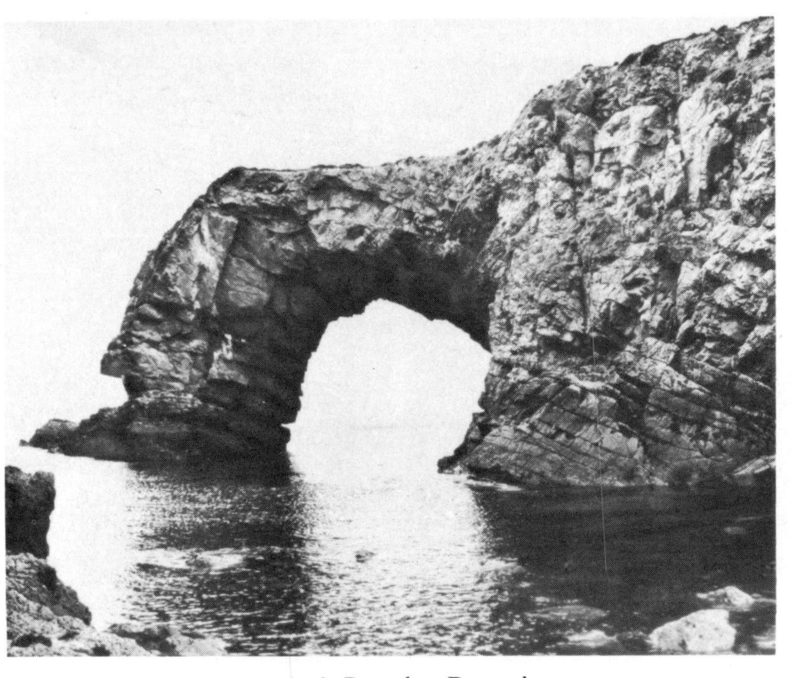

41. Arch, Portsalon, Donegal

42. Stack, shore of L. Huron, Ontario

43. Cliffed face of Slieve League, Donegal

44. Offshore bar, Long Island, U.S.A.

45. Groynes, Shoreham

46. Small delta in flooded sandpit

47. Vertical view of the Colorado Delta

48. Drowned valley, South Devon

49. Raised beach, Islay, Argyll

50. Raised-beach deposits, Guernsey

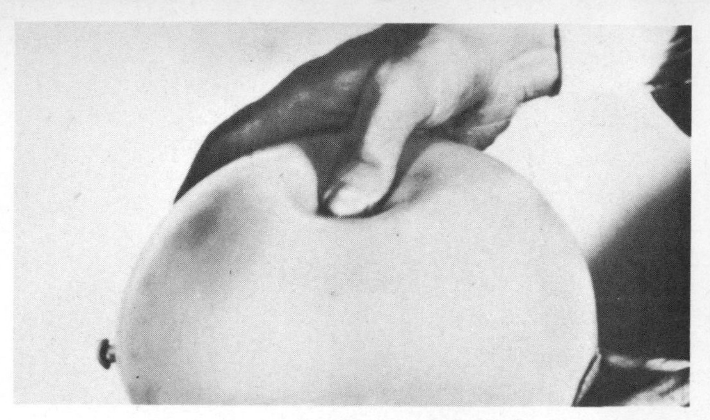

51. Depression in a balloon

52. Depression in a balloon coated with hinged plates

convenient point of reference for dating the higher platforms of the west.

This brings us to the general question of dating platforms, as opposed to the dating of shorelines. As pointed out for Wales, a high and underformed platform implies a high stand of the sea, even though the platform was developed on land, and the former shoreline has vanished. Several approaches to dating are possible. All involve reference to the geological scale.

Planation of a landmass must involve deposition of sediment in the sea. It seems a reasonable proposition that initial uplift, by triggering vigorous erosion, would promote equally vigorous sedimentation offshore. As planation became more and more advanced, the supply of sediment should dwindle. Destruction of the landmass means elimination of the source of sediment supply. Renewed uplift to any considerable degree would reactivate the supply of sediment. In this way, the offshore record of sedimentation ought to provide a kind of mirror-image of the course of denudation on land. But in taking this particular approach, we should by definition be dealing with areas of crustal instability. Furthermore, not very much is known of the history of sedimentation during the last 50 million years or so, except in respect of certain great deltas, those of the Mississippi and the Rhine included among them. Deltas do in fact collect a large fraction of the sediment which is now coming off the landmasses; but the record integrates the whole supply system for a given river basin. In any event, the matching of erosional with depositional history might be possible only for the greatest of all events of planation.

The study of sediments on land is in some senses easier. In South America, North America, Africa, and Australia exist large, shallow, sediment-filled terrestrial basins, or spreads of terrestrial sediment distributed on plains below a mountain front. Some of the sedimentary units apparently belong within the last 50 million years. Where they rest on planated surfaces, they supply a limiting age for the operation of planation – if they themselves can be dated.

Unfortunately for hope in this direction, the deposits in question are typically poor in fossils, and frequently devoid of fossils altogether. Where fossils do occur, they tend to be poorly preserved, and in addition to offer little help in fixing age. For a given large area, even the relative dating of deposits separated from one another in space can be a matter of guesswork. Fossils of land organisms are rarely abundant, except for plant fossils in a few localities. But plants have evolved far less rapidly than have some organisms in the sea. There is nothing on land to match the delicate record of marine creatures, which evolved fast and spread widely, permitting detailed and accurate correlation between area and area, and detailed and accurate subdivision of the geological column.

*

At the other extreme from those workers on platforms who infer former sea-stands at very high levels, some writers have demanded former stands at very low levels – as much as 4000 feet (1250m)

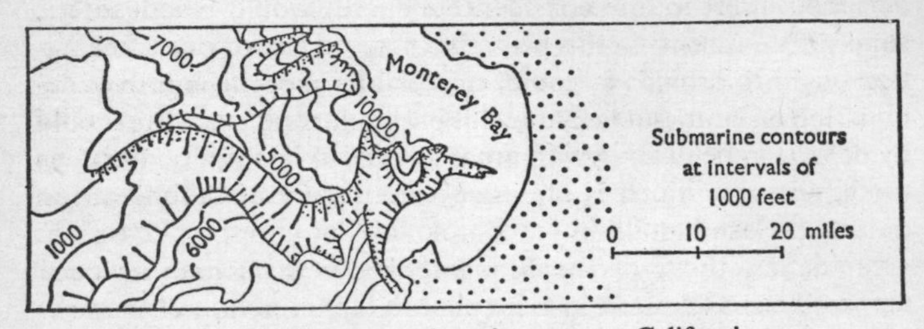


Fig. 60 Monterey submarine canyon, California

below the existing mark. Their opinions have now become superseded by others. The actual evidence remains undisputed. It is provided by submarine canyons, enormous valleys which gash the continental slope (Fig. 60). This slope leads down from the shallow offshore belt to the deep ocean floor.

It was originally thought that the canyons continue the courses of large rivers across the sea-bed. Some do precisely that – the canyons off the mouths of the Hudson, the Congo, and the Indus run seaward from the present river-mouths. There is little doubt that this fact has led many commentators to believe that the submarine valleys are nothing more than submerged continuations of river-valleys on land. But since the canyons descend to depths of thousands of feet, they can only have been cut by rivers if the sea was at one time thousands of feet lower than it now is. No mechanisms can be imagined capable of lowering sea-level by the enormous distances required.

Very many canyons, however, begin far out to sea, characteristically near the top of the continental slope. The most likely explanation seems to be that they are cut by turbidity-currents. A turbidity-current can flow for very long distances when its specific gravity exceeds that of the surrounding water. High-density currents can carry very heavy loads in suspension, and are capable of transporting large fragments of broken rock. Turbidity-currents can probably be set off under water if mud slumps on a slope of 2° or even of 1°. The slumped mass, becoming diluted with water, forms a dense cloud which moves rapidly downslope. Deposition of mud off the mouths of large rivers, disturbance by earthquakes, and the motion of violent sea-waves produced by earthquakes, have all been advocated as possible causes of submarine turbidity-currents. Such currents could incidentally provide a convenient explanation of the arrangement of thick formations of fine sediment in very thin layers. They might well be capable of eroding canyons no smaller, and no less impressive, than the Grand Canyon of the Colorado.

Where canyons do come close inshore, they can serve as drains for beach sand. Some do just this on the California shoreline. Many beaches, deprived of part of their sand supply by the damming of rivers inland, but still draining away to the deep ocean, are perceptibly dwindling. Some may be threatened with disappearance. The inevitable result will be increased shoreline erosion.

CHAPTER ELEVEN

# *Ups and Downs of the Land*

EROSION of the land cannot be properly discussed without some direct or implicit reference to uplift. It has already been necessary to mention repeated uplifts in certain areas; to point out that sedimentary rocks, formed beneath the sea, have been elevated to their present positions; and to say that regional uplift initiates the cycle of erosion. On a later occasion, it will be noted that parts of the earth's crust have been depressed beneath loads of ice, and that those areas from which the ice-caps have melted away are now rising (p. 160). It is time for something to be said about the mechanism of elevation and subsidence.

Geology supplies abundant and indisputable evidence that uplift and depression, during the last 500 million years, have been affecting continental masses of the kind we know today. Uplift is recorded by gaps in the geological record – the results of erosion, when uplifted rocks were destroyed. Depression is shown by marine sediment resting on severely eroded surfaces, which must have been submerged in order to be covered by the rock-waste accumulating beneath the sea. On the scale of geological time, upward and downward movements of the land seem rapid. They are also notably irregular, varying in intensity, in distribution, and in direction.

Three topics are particularly important in this context. It has to be explained why ice-caps, enduring for geologically short periods, can push down those portions of the crust on which they rest. It has to be shown why some land-areas reach heights of many thousands of feet – in other words, why huge mountain-belts can be supported while ice-caps cannot. Finally, alternating elevation and depression in the same area have to be accounted for.

Although details of alternating elevation and depression on the local and regional scales have still to be worked out, the general picture is clear. It portrays a rigid and thin superficial crust which deforms mainly by breaking, underlain by material which can behave in more than one fashion, according to circumstances. As will be seen, the reconstructed history of deformation goes back far into time. But the response to loading by heavy caps of ice, and the record of this response, amount to a series of experiments in the short geological term.

In the colloquial sense of the word, the superficial crust is manifestly rigid. It breaks along lines of faulting. The most rigid blocks of all consist of extremely ancient, greatly altered, and almost exclusively crystalline rocks. But even these are known to have yielded under the weight of ice-caps – as in the Baltic area and the Canadian north. Since it is inconceivable that a very thick layer of rocks similar to those exposed at the surface could yield freely, it is necessary to infer that weaker material is present at no great depth.

The idea that a rigid crust can be gently depressed under load is not so paradoxical as it seems. The rise which followed unloading did not increase smoothly towards the central areas, where the load was greatest, but varied abruptly across certain lines. A simple illustration will clarify this statement. When a depression is made in an elastic surface – as when a thumb is gently pressed against an inflated balloon – the sides of the depression curve smoothly inwards (Plate 51). But imagine the balloon to be entirely covered with hinged plates; the sides of the depression then become angular in profile, and tilting occurs along the lines of the hinges (Plate 52). Land-areas which were pushed down by the weight of ice-caps, and which are now rising, are moving precisely as if they were so hinged – indeed, the term *hinge-line* is applied to those lines where the rate of uplift changes sharply.

Until recently, it was unknown whether or not the loads formerly imposed on the crust by great ice-caps were fully compensated by subsidence. Observations in Greenland (p. 156) show that compensation there is complete. The crust is evidently very sensitive to unusual loading. The weight of sediment in the Mississippi Delta, too, is thought to be fully compensated by subsidence. Conversely, it can be inferred that the crust is equally sensitive to unloading. Uplift is now going on where the great ice-caps used to stand.

In addition to being strong, the crust of the earth must also be light. The world's lands stand high simply because they consist of relatively light rocks. They are buoyed up by the heavier but weaker material beneath. The relevant principle is that of equipoise – or, to give it the usual name, the principle of *isostasy*.

Reduced to the component elements, the rocks of the world's lands are chemically similar to granite. Some 65 per cent of the whole consists of the compound silica ($SiO_2$), and another 15 per cent of alumina ($Al_2O_3$) – a total of 80 per cent. Hence the collective name *sial* – si for silica and al for alumina – given to the light rocks of the continental crust. The sial has a density of about 2·80.

Beneath the great oceans denser rocks occur. These, resembling basalt in their bulk composition, contain high proportions of silica and magnesia (MgO) – hence the name *sima*. Part of the sima, with a density of about 3·0, is crystalline. The crystalline sima and the sial combine to form the earth's crust, about 40 miles thick (Fig. 61).

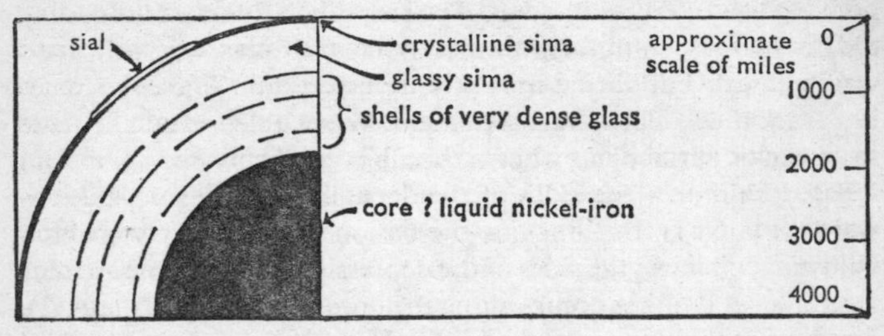


Fig. 61 Earth-shells

Beneath the crust lies sima which has not yet frozen. Despite its high temperatures (p. 49), it is kept rigid under pressure. The light continental blocks, buoyed up by the non-crystalline sima, have frequently been compared to icebergs floating in the sea. But apart from the fact that one-ninth of the continental masses projects above the general level – a faction observed also with icebergs – the comparison has little to commend it. The non-crystalline sima is not liquid but glassy.

At atmospheric pressure and normal temperatures, glass will flow, bend, and break. Under sudden strain it is shattered. Within limits, a large sheet can be flexed. But under its own weight, ordinary window-glass very slowly thins at the top of the pane and thickens at the bottom. In breaking, it acts as if it were a solid. In bending and recovering, it displays elasticity. In slowly flowing, it is plastic. A substance which flows, bends, or breaks, according to the conditions to which it is subjected, is a *rheid*. The glassy sima beneath the rigid crust appears to be in a rheid state.

Response to strain on the part of a rheid is governed partly by the amount of strain and partly by the time-factor, as can be seen from the comments made on the behaviour of window-glass. A parallel set of illustrations is given by the behaviour of the silicone familiarly called bouncing putty. A small body of this entrancing material can be moulded into a ball – i.e. plastically deformed. Thrown down, the ball bounces – it acts elastically. It can easily be wrenched apart, breaking across as if it were a solid. Drawn out like a rope, the putty sags in the middle to form a festoon – that is, laminar flow occurs. The amount of strain needed to produce a given deformation increases from bouncing putty to window-glass, and from window-glass to the glassy sima. But because very great strain is in fact applied, the rate of movement in the glassy sima is noticeable. Although direct measurement is not possible, reasonable estimates are. They run at about 20cm a year.

Loading by ice, and the duration of loading, were evidently enough to cause movement in the glassy sima, and to bring about isostatic depression. This being so, it must be concluded not only that the continental blocks in general are light, but also that the rocks of mountain-belts are particularly light. Unless the great bulk of rocks in mountain-chains were compensated in some way, isostatic subsidence would offset any tendency for mountainous areas to rise. In actuality, the common history of mountain-belts is one of general and repeated uplift. It follows, therefore, that the visible bulk of mountains above the general level is offset by much greater bulks of light rock below that level. Mountain-belts have light roots striking deep into the body of the earth.

The principle of isostasy was, in fact, originally formulated with reference to mountain-belts. Direct evidence that high relief is in some way counterbalanced first became widely known about 125 years ago. During the survey of India, it was found that the Himalayas fail to exert the gravitational pull which would be expected from their size. Extreme suggestions – including one that mountains are hollow – were soon abandoned in favour of the idea that mountain-belts are supported by unusually light material at depth – that is, by masses of sial projecting down into the glassy sima.

Ignoring for the time being the way in which mountain-roots are formed, we may note that they can be held responsible for the first general rise of a mountain-belt and also for subsequent rises. Even if a mountainous region is originally in perfect isostatic equilibrium, the balance is bound to be destroyed by erosion. As landslides and torrents tear the mountains away, and as the main rivers carry debris down to the sea, the load is reduced. Continued intermittent uplift is therefore to be expected. It can be argued that the Himalayas, continually reduced in bulk by denudation, are still rising. Their exceptional height may be no more than the consequence of deep valley-cutting. Although their *average* height is decreasing, the individual heights of single peaks may well be on the increase.

*

The concept of the earth as composed of concentric shells comes from the study of earthquake-waves. Three kinds of wave are generated by an earthquake. Long waves (L) travel along the surface. Both push waves (P) and shake waves (S) go underground. Both types are subject to sudden bending – refraction – where they cross a boundary at which speed of transmission changes abruptly. Analysis of the depths at which bending takes place defines concentrations at particular depths. These are the depths which separate one earth shell from another. Contrasted behaviour of the P and S waves when they enter the earth's core lead to the inference that this core is liquid, although the liquid condition is perhaps very different from anything we know at the earth's surface in ordinary circumstances of atmospheric pressure.

The two outermost shells are moving. These are the discontinous shell of sial, and the continuous shell of crystalline sima. The rate of movement varies from zero to about 2cm per year. Its impulse is either push or pull, depending on one's view of the relationship of strain on the crust to strain on the underlying mantle. Recognition and measurement of movement in the crust have completely revolutionized geological science. They concern the chief locations of earthquakes, the distribution of volcanoes, and the sundering of whole continents.

As soon as the borders of the Atlantic Ocean had been mapped, sundry observers began pointing out that Africa and South America can be fitted together, like pieces in a jigsaw puzzle. The fit improves when edges are drawn, not along existing shorelines, but along the edge of the continental shelf. This is where the blocks of continental sial terminate. In the early years of this century, Alfred Wegener proposed an hypothesis of continental drift. His investigations of the distribution of ancient climatic belts led him to reconstruct a supercontinent, Pangea, for a time about 250 million years ago (Fig. 62). The jigsaw includes not only Africa and South America, but other major landmasses also.

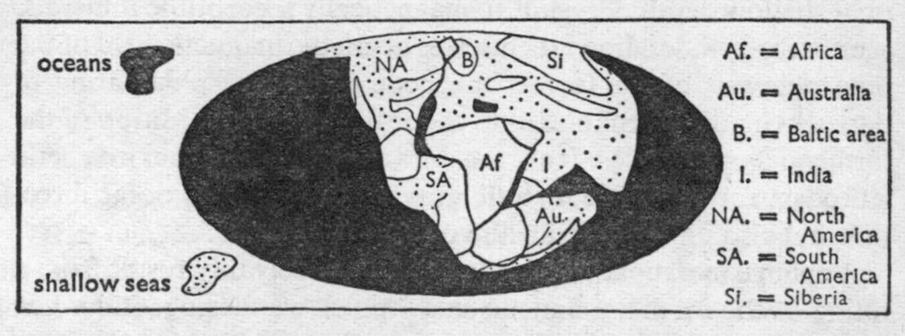


Fig. 62 Suggested distribution of lands 250 million years ago (adapted from Wegener)

The break-up of Pangea was ascribed by Wegener to the process of continental drift – the radial movement of continental masses away from the poles. The drift hypothesis attracted much attention, long mixed liberally with abuse. Wegener seems to have been resented by the geological fraternity for having been chiefly a meteorologist, and thus apparently unqualified to deal in geological matters. In addition, the idea of drift was vigorously opposed by those who claimed that no driving mechanism could be identified. It seems ironical that the evidence of past glaciation should have become accepted at face value, even though the causes of ice ages, and of fluctuations within ice ages, are still not understood today, while evidence for the former linkage between continents should have been rejected by all but a few. Among these few was the South African geologist du Toit, who found himself convinced by the matching distributions of rocks among the landmasses of the southern hemisphere. The question is more than one of general similarity. On either side of the south Atlantic, for instance, the distributions of rock types fit into the jigsaw, just as clearly as do the edges of the continental shelves.

Beginning in the early 1960s, views on continental drift changed very fast indeed. The new contributions came from several directions. The earliest related to paleomagnetism. This, the study of ancient magnetic directions, shows that the magnetic poles have varied through geologic time, in their relationship to the major landmasses. Former magnetic directions are preserved in many rocks – particularly in crystalline rocks which froze at the surface or at shallow depth. Crystals of magnetically susceptible minerals, such as iron oxides, lined themselves up on the magnetic field of the time. Analysis of the former magnetic directions, plus dating of the rocks involved, enables us to define the apparent position of the former magnetic poles. It also enables us to define former magnetic latitudes; for magnetic direction involves not only a polar direction, but also a magnetic dip into the earth's interior.

A change in magnetic direction could conceivably result from a tilting of the earth's magnetic axis – polar wandering. But when directions are determined for a given time in the past, and for several landmasses at that time, the determinations only make sense if the landmasses are inferred to have moved differentially. Also, the determinations of former latitudes require differential movement. When the landmasses are rearranged, for any time up to 250 million years ago, so as to allow for their former orientations and former latitudes, the result is the reconstruction of a supercontinent, Wegener's Pangea, either intact at the beginning, or in various stages of break-up subsequently. There is one important elaboration, however. In the early reconstructions, the existing sites of the mountain belt running from Tibet and the Himalayas into Turkey are occupied by the basin of the Tethys Sea. This sea separated Pangea into two supercontinents, Laurasia in the north and Gondwanaland in the south. The existing Mediterranean is its last remnant, doomed to be closed in as Africa drives northward towards Europe.

The most rapidly-moving landmass is probably India, with Australia running a close second. Rates of movement run at about 1 or 2 cm/year, one-tenth of the rate estimated for the underlying glassy sima. But 2 cm/year for 250 million years amounts to a half-million kilometres. The Indian block has gone far enough, and fast enough, to wedge a large part of itself under Asia. The under-driven portion provides the light support required to sustain the heights of the Himalayas and Tibet.

The second line of evidence takes us to mid-ocean. Early explorations of the deep ocean floor identified mid-ocean rises. On the basis of occasional point soundings by weighted line, these rises seemed to have a total relief comparable to the relief of mountain-belts on land, but to be generally smooth. The details supplied by echo-sounding not only revealed distinctly rugged forms, but also showed that a typical mid-ocean rise is split down the middle, as if it were rifted. So it is.

At the mid-ocean rift, the rate of escape of heat from the earth's interior is unusually great. We infer that the heat is being brought up by convection currents in the mantle, the glassy sima of Fig. 61. The inference of convection is enormously fortified by the fact that mid-ocean rifts can be shown to be widening. As they widen, molten sima from below freezes on their sides. It provides a continuous record of formation of the ocean-floor crust.

The paleomagnetic evidence mentioned above includes not only signs of change in direction, but also of reversal of direction. From time to time, the earth's electromagnetic system flips over. What we now recognize as the north magnetic pole becomes the south magnetic pole. The reversals are faithfully recorded in the slabs of new crust which are formed at the margins of the mid-ocean rifts The record on one side is a mirror-image of the record on the other (Fig. 63).

The third line of evidence relates to the distribution of earthquakes, volcanoes, and mountain-belts. In the gross view, the major belts of young fold mountains are two, one encircling the Pacific, the other running across Eurasia from the Mediterranean to China. These are also belts where many volcanoes are concentrated, and where many earthquakes happen. But volcanoes and earthquakes also appear in mid-ocean. When their distribution is mapped, they define the boundaries of major crustal plates. The new science of plate tectonics deals with the relationship of plates to one another.

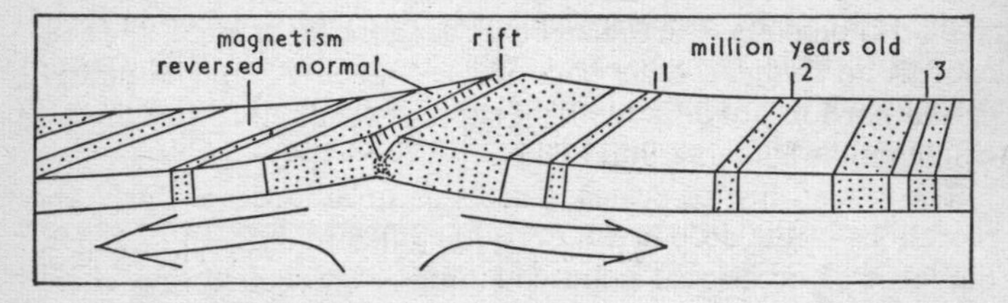


Fig. 63 Mid-ocean ridge, showing the mirror-image record of magnetic reversals

Plate tectonics makes abundant sense out of volcanoes, earthquakes, mountain ranges, and mountain roots. It provides the starting-point for the work of landmass destruction. It teaches that the earth's crust is divided into some twelve major fragments; and that these fragments are renewed in mid-ocean and consumed at the continental margins (Fig. 64).

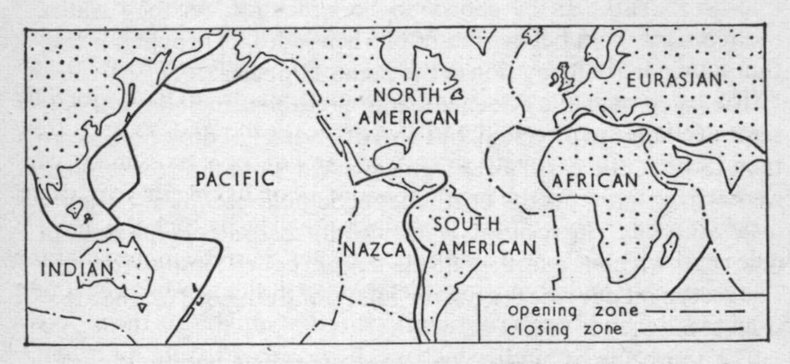


Fig. 64 Sample of plate distribution, simplified, showing zones of opening and zones of closing

What comes up must go down. The crust that is formed at the edges of the mid-ocean rifts will eventually reach the continental margin, will subside, and will be re-melted. The belt where subsidence and re-melting occur is the *subduction zone* (Fig. 65), where oceanic crust is reabsorbed. But the subduction process appears capable of generating a great deal of heat. In particular, it can melt enough subsurface material to supply emplacements of granite (sialic in gross chemistry), and to drive the volcanoes of the Pacific ring of fire. Also, sudden release of strain in the subduction zone is a highly probable explanation of many earthquakes. The foci of deep-seated shocks typically slant obliquely down, where an oceanic plate is plunging under a continental plate.

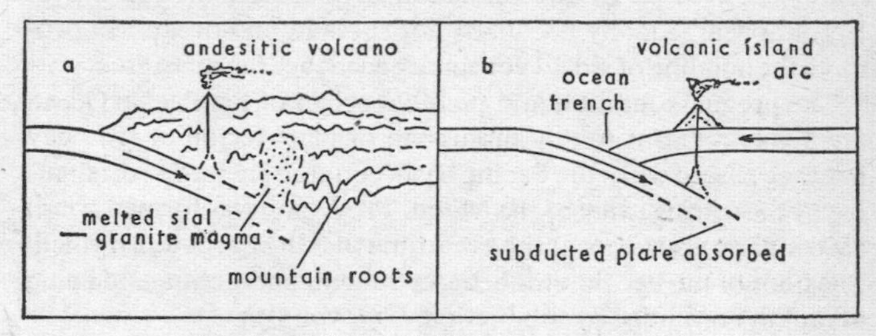


Fig. 65 Subduction (*a*) at continental margin
(*b*) in mid-ocean

Views of the subduction process are mainly two. These two are to some extent diametrically opposed. On one view, the mid-ocean rifts are wrenched apart by currents in the mantle. These, rising by convection, come near the surface beneath the mid-ocean ridges, and are forced to spread sideways. The sideways movement, estimated (as previously noticed) at about 10 or 20 cm per year, drags the overlying crust apart. The opposite view is that the surface crust, simply because it is frozen, is denser than the material of the immediately underlying mantle. It sinks in the subduction zones, pulling the mid-ocean rifts apart as it does so. Here is the proposition, either push or pull, which was heralded earlier in this chapter.

Either view is perfectly compatible with the known distribution of earthquakes and of volcanoes of all types. Many central volcanoes occur above subduction zones, where the reabsorption of oceanic crust into the mantle releases much heat. Part of this heat is used in the melting of sialic crust, to produce granitic magma. Eventually this magma will freeze deep underground, in the form of new granite bodies. So deeply insulated is the magma by overlying rock, that it can need 50 million years to cool off.

Central volcanoes above subduction zones typically erupt lava of mixed composition. This lava, characteristic of the entire Pacific rim, is called *andesite*, after the Andean volcanoes of South America. The volcanoes in question are all cone-builders, given to violent activity and to the emission of vast amounts of dust. When the sialic crust is ripped open, and liquefied material from the mantle is released by the reduction of pressure, then floods of basalt appear. Finally, localized hot spots in mid-ocean can promote the building of shield volcanoes, as in the Hawaiian area.

The present condition and possible origin of the Pacific Ocean are highly relevant to any discussion of plate tectonics. The day has long passed when the Pacific basin could be guessed as originating by the tearing away of the Moon. The Moon was formed about 4500 million years ago, at the same time that the Earth was formed. The floor of the Pacific, still being created in mid-ocean, and being as rapidly consumed by subduction, is far younger.

However, the floor of the Pacific is special in several ways. It contains the Hawaiian hot spot, as yet the best studied hot spot of all. It also displays the only case in which a mid-ocean ridges comes ashore. The motion involved here is side-to-side slip, rather than a tearing apart. The Pacific plate is sliding northward relative to the North American plate. A more accurate statement might be that the North American plate is rotating in a counter-clockwise direction. Here is the origin of the San Andreas and related fault systems in California. The seaward slice of crust belongs properly, not to the continent but to the ocean.

Recognition of subduction zones in general has greatly simplified the problem of accounting for belts of fold mountains. When these were thought to result from the vice-like squeezing of vast wedges of sediment, two sides of the vice had to be provided. The only possibility seemed to be two large rafts of sialic crust. In this way, fold belts such as that of the Appalachians were thought to demand former continents where oceans now occur. One particular demand, linked in pseudo-science with mythology, is that for a lost continent, Atlantis. But sialic crust, being by definition light, cannot simply founder and disappear. The Appalachians were crushed into concertina folds, at a time when the Atlantic Ocean had not yet formed by the sundering of Laurasia. In contrast, the cordilleran belt on the western margins of the Americas relates to the underdrive of the Pacific floor.

An obvious candidate for disruption might seem to be the African landmass. Drift has already opened the Jordan Valley rift (Plate 36), and has produced the Red Sea, including the northern branches. Obscured by volcanic deposits in parts of Ethiopia, the rift system runs far to the south. It contains the long, narrow, deep lakes of eastern Africa. However, although the southern rifts contain a few uneasy volcanoes, and are certainly associated with tilting of the crust (Fig. 30), the splits show little tendency to widen further – at least for the time being.

When we look far back into distant geological time, urgent questions appear. Answers promise to be extraordinary. Wegener and his colleagues took the time of about 250 million years ago as a starting-point, because the relevant distributions of glacial, arid, and coal-producing environments seemed to fall into a recognizable pattern. Extensive glaciation in the southern continents can indeed be accounted for by the reassembly of Gondwanaland. Deserts pose problems. We know little or nothing of the general atmospheric circulation, even as late as 5 million years ago. And deserts can be produced almost anywhere in the world, if mountain-belts exist to wring the moisture out of rain-bearing winds. Coal measures also misled many researchers in the early days. It is now clear that they come in two sorts – those formed in the humid tropics, and those formed in temperate midlatitudes.

Nonetheless, there is now very good agreement on the existence of Pangea (or of Laurasia and Gondwanaland plus Tethys) at the indicated time. The history of break-up, and the paths of movement of individual plates, are reasonably well defined. But if Pangea occupied one side of the globe, what was on the other? Also, what arrangement of the continental blocks preceded the arrangement in Pangea?

Evidence which is as yet scattered suggests that plate movement operated far back in geological time. After all, we have some 3500 million years to deal with. In some places there have been identified sutures, where two moving blocks became welded together. It is possible that the clumping of sialic blocks illustrated by Pangea, and the subsequent break-up, had precursors in the interval between 3500 million and 250 million years ago.

But just as the Atlantic, Indian, and Antarctic Oceans can be closed by the reassembly of continental blocks, so can the Pacific Ocean. If the globe is shrunk enough to eliminate the ocean areas, *all* the continental blocks fit together. The inevitable suggestion here is that the globe has greatly expanded, since about 250 million years ago. Such a suggestion is terrifying. If it is valid, expansion is presumably still going on. Its ultimate result could be the planet.

Opposing the idea of dramatic expansion is the record, very far back in time, of sedimentation in the sea. It is in the ocean that life began. Colonization of the lands by organisms came quite late in the piece. Moreover, the similarity between the west of North America and the east of Asia could be due simply to the drift of a former portion of Asia, across what is now the North Pacific, and its attachment to a former, and smaller, North American land-mass.

Much, obviously, remains to be learned. Particularly is this true for the cellular patterns of circulation in the mantle. If the number of cells changes, so must the movements of plates. Here is a possible explanation of bursts of mountain-building and of episodes of extensive submergence, such as are well documented in the actual record.

CHAPTER TWELVE

# Ice in the Mountains

GLACIAL and arid conditions have aptly been styled climatic accidents. On the geological time-scale, glacial erosion operates briefly and seldom – perhaps for about one-half per cent of the time. Although deserts may be less temporary than glaciers, there is evidence to show that desert margins fluctuate considerably. Thus, although schemes can be drawn up for the cycle of erosion in very cold or in very dry climates, it is very doubtful if a glacial cycle could, in practice, run its whole course, and quite doubtful if arid cycles are likely to be completed. Glacial and desert processes are therefore regarded as likely to modify, but not necessarily to transform, landscapes which have been modelled by the processes of normal erosion.

Accounts of glacial sculpture usually contrast highland with lowland glaciation, or valley-glaciers with ice-sheets. There is much to be said for taking valley-glaciers on their own. They illustrate in the clearest possible manner the ways in which surface features can be altered during glaciation. Nevertheless, it should be remembered that the two contrasted classes, valley-glaciers and ice-caps, merge into each other where ice-caps discharge tongues of ice into valleys or where highland is wholly covered by sheet-ice.

Glaciated highland is spectacular because ice is highly erosive and slow to move. Ice-tongues, shod with angular rock-fragments and made abrasive by a charge of sharp sand, rasp the beds and side of their channels. Slow-moving ice needs larger channels than those of the streams which it temporarily supersedes. The huge trenches which are often called glaciated valleys are really the beds of former glaciers. They are best called *glacier troughs* (Plate 53).

Rates of movement vary greatly, both from glacier to glacier and from one part to another of a single glacier. They also vary from time to time at a given spot, in accordance with the amount of ice coming down. This amount depends, in turn, on the balance between snowfall and melting. Movement is slight in the catchment area, becoming greater at lower altitudes but decreasing again towards the snout. Kuenen compares the speed of moderate-sized Alpine glaciers, in their fastest portions, to the movement of the hour-hand of a watch – about 150 feet a year. The Rhône Glacier moves about twice as fast as this, and the Great Aletsch Glacier nearly four times as fast, advancing 550 feet in a year. Glaciers in Central Asia, possessing catchments of 400 square miles and reaching lengths of 40 miles, cover 2500 feet (half a mile) in a year. Much higher speeds are recorded on the tongues of ice which, draining from the Greenland cap, travel 80 feet in a day and 5 miles in a year. But even at this pace, supposing it were maintained all the way out from the centre, the ice now melting on the margins of Greenland or breaking away as icebergs may be 10,000 years old. Antarctic bergs conceivably include ice formed 20,000 years ago.

Moving at half a mile a year, a valley-glacier is bound to need a channel of great dimensions. The features characteristic of such a channel are exposed both near the ends of existing glaciers and in highland areas which carried glaciers during the cold episodes of the Ice Age. Near the ends of many present-day glaciers, erosional forms have recently been revealed and depositional forms recently created in the course of a general recession of the ice. Elsewhere, former glaciers having disappeared entirely, the whole range of sub-glacial features is revealed. The present recession is a minor event, even on the short time-scale of the Ice Age, but it does serve to show that glaciers the world over respond simultaneously to changing influences.

Thorarinsson has concluded that the Icelandic glaciers were considerably less extensive from 900 (when the island was first colonized) up to about 1350 than they were in 1800. Soon after 1700 began a general advance, which reached its maximum about 1750. During the next half-century they were either stable or receding, but a re-advance started in the early 1800s and reached its maximum about 1850. Stagnation or recession followed once more, with a slight halt in the early 1880s and a subsequent general retreat. Scandinavian glaciers are thought to have behaved similarly. Retreat from the lines reached in the late nineteenth century is confirmed from many parts of the world.

Between the late nineteenth century and the year 1948, the Columbia Glacier in the Canadian Rockies receded three-quarters of a mile. In about 30 years, the lowest 3½ miles of the glacier lost 350 million cubic feet in volume. The snout of the Athabaska Glacier went back nearly half a mile in 40 years. The Saskatchewan Glacier has lost two-thirds of a mile since the 1890s, and receded no less than 250 feet between 1945 and 1947.

The general picture is one of glaciers now rapidly receding, and of considerable fluctuations of their fronts in periods as short as a century or even half a century. Small wonder that formerly glaciated highland areas are littered with the signs of intermittent retreat, in the form of moraines laid down wherever the glacier snouts were stationary for a time.

As used above, the word *retreat* must not be taken to imply a withdrawal of the ice-tongues. It merely signifies recession of the front, caused by excess melting. For all their slowness of movement, valley-glaciers descend far below the snowline. They end where melting is powerful enough to destroy them. Advancing snouts correspond to increasingly cool climate, to increased snowfall in the catchment area, or to both combined. Retreating snouts correspond to increasingly warm climate, to decreased snowfall, or to both. Stable snouts show that, for the time being, supply of ice counterbalances melting. But because ice in a big glacier takes a long time to travel from the catchment area to the snout, the effects of minor changes in temperature and those of changes in snowfall can obviously be out of phase. In the early years of this century, despite the general tendency of glaciers to recede, some tongues advanced into growing forests.

Relationships of valley-glaciers are best displayed where mountain-tops rise bare above the glacier-heads (Plate 54). As a rule, upstanding peaks are jagged in outline and precipitous on the flanks. The much-photographed Matterhorn is accepted as representing the type. A whole field of peaks corresponds accurately to the popular stereotype of mountain scenery.

Pyramidal peaks are but remnants of former mountains. Their sheer walls consist of the steep slopes developed at the heads of glacier-troughs. Where erosion has been rather less severe, and where the ice has vanished, distinctive hollows are displayed

(Plate 55). These are *corries*, the essential land-forms of glaciated highland.

Corries originate as ordinary valley-heads. With the onset of glaciation, they are occupied by snow. Settling under its own weight, the snow loses part of its contained air and changes into *firn*, a substance intermediate between snow and glacier ice. New snow becomes firn in the course of a few months. Compressed by additional falls, firn passes into ice. In climates like those of the Swiss Alps, a thickness of 100 or 150 feet is enough to set the ice in motion.

In corries which still contain ice, it is usual to find a deep crack – the *bergschrund* – running round the margin of the ice close to the enclosing wall of rock (Plate 56). The crack is probably due partly to downhill movement of the ice, and partly to local melting at the contact with the rock-face. Water running into the crack, and freezing in fissures in the rock, causes shattering so powerful that the wall retreats like a steep cliff. Broken fragments are frozen into

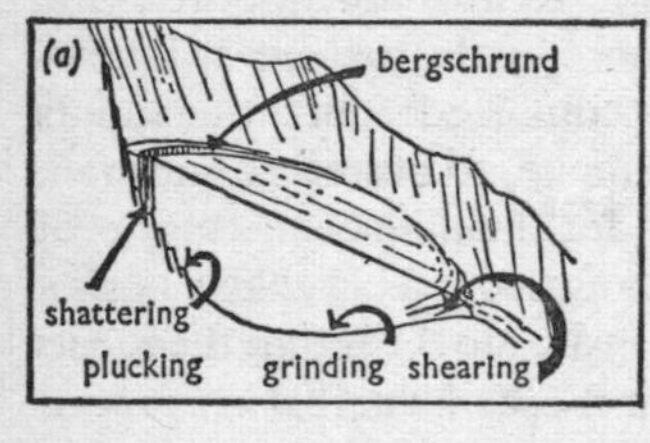


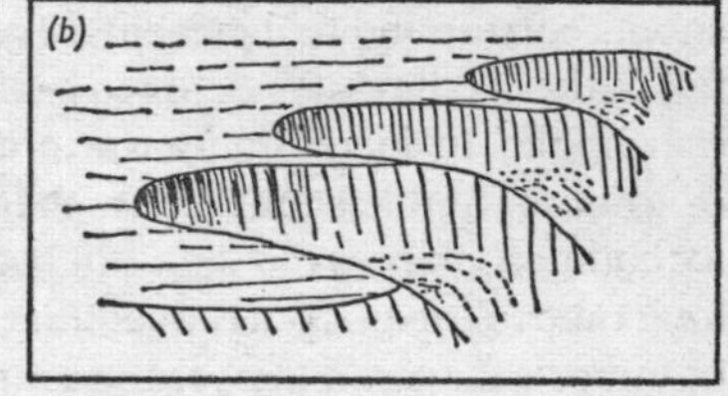


Fig. 66 (*a*) Corrie-cutting
(*b*) Corries at the edge of a plateau

the ice and carried away. In this way the hollow is extended headwards and sideways. By another process it is deepened. Snow, firn and ice, accumulating during the winter, become mechanically unstable. As the whole mass rotates, the base of the hollow is ground down (Fig. 66*a*). If the central floor is reduced below the level of the lip, a lake forms when deglaciation occurs and the ice melts away.

Corries scalloped in the side of a plateau look as if they have been punched out by a gigantic pastry-cutter (Fig. 66*b*). Narrower belts of high ground are frequently attacked more severely on one side than on the other – in Britain, northern and eastern flanks have suffered most. A single peak, savagely bitten by three or four corries, is reduced to the pyramidal form so well exemplified by the Matterhorn. At a slightly earlier stage in destruction, the walls of adjacent corries merge in sharp ridges which are kept continually trimmed by frost (Plate 56).

Similar processes to those at work in corries are thought to operate where valley-glaciers cross bands of especially resistant rock. Such bands are likely to have produced waterfalls and cascades on pre-glacial mountain streams. Glaciers moving down the former stream-valleys descend in ice-falls, below which rotational slipping may go on. Again, air of varying temperature and melt-water released during summer can both penetrate the cracks, favouring frost-action on the underlying rock. It is therefore possible for steep valley-reaches to be further steepened during glaciation, so that great steps are exposed at glacial retreat.

Visible rock-steps have been scoured into smoothness on their up-valley sides, and plucked or shattered into roughness on their down-valley sides. Isolated rock-knobs have been similarly affected (Plate 57). They supply undeniable – and invaluable – evidence of the direction taken by the basal ice. Rock-knobs in numerous groups are typical of sheet glaciation rather than of erosion by valley-glaciers; but they deserve a mention, if only for their unforgettable resemblance to schools of surfacing porpoises. Examples occur in Anglesey and on the Swedish island of Hönö.

Rock-steps and rock-knobs are among the minor irregularities which break the profiles of glacier-troughs. The largest irregularities consist in elongated lake-basins, some of which attain very great lengths and are eroded hundreds of feet below the levels of the sills at their lower ends. Lake Okanagan, in British Columbia, is no less than 70 miles long. The floor of the Scottish Loch Morar descends to 1000 feet below sea-level. Still greater depths are known where troughs open directly into the sea as fiords – soundings of 4000 feet have been made in the Sogne Fiord, and one of 4250 feet is recorded from the Messier Channel in Patagonia.

As was previously said, a large glacier can come far below the snowline before melting completely away. Now at the headward end, where minor glaciers move down minor valleys, the depth of ice in any one valley is relatively small. Towards the snout, the main tongue is progressively reduced in volume and in thickness, its melt-water trickling away to feed the stream lower down. In the middle portion, the trunk glacier is at its thickest and most vigorous. So long as its axis of gravity slopes down-valley, the glacier can move (Fig. 67). As it is thicker in the middle reaches than at either end, its base can descend below the horizontal, and a long basin can be rasped out.

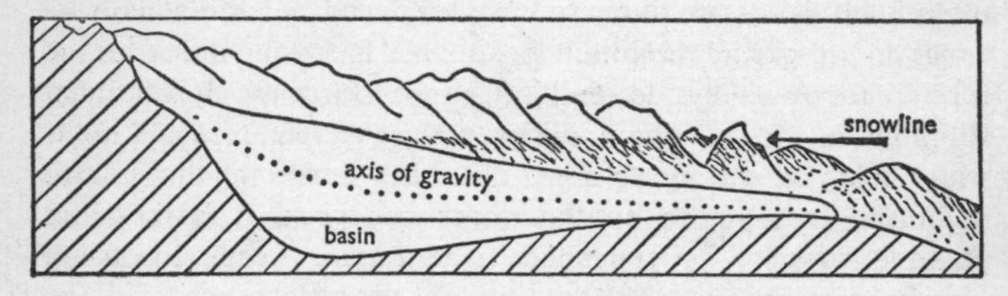

Fig. 67 Basin-cutting by a valley-glacier

The detailed mechanism of glacier flow continues under investigation. Some authorities object strongly to the description of a glacier as a river of ice, maintaining that the term incorrectly suggests that glacier movement resembles the flow of water. It is true that valley-glaciers, like very many rivers, move most rapidly near the centre and at or near the surface. But ice is unable to undergo deformation as rapidly as water can. Both for this reason, and also because ice acts to some extent as if it were rigid, the shear on the bed and banks of a glacier channel is often far greater than that on the bed and banks of a stream channel.

We make an exception in respect of the glaciers of high latitudes and of some high-altitude areas: these, the so-called cold glaciers, are typically frozen to their beds. With so-called temperate glaciers, which are not so frozen, the speed of basal slip can be more than half the maximum speed of movement at the surface. We are thinking of slip rates of the order, say, of 15 to 25 m per year. Slip rates increase on steep slopes, where temperate glaciers can be highly abrasive. If meltwater is supplied, the slip rate increases again. Lubrication by meltwater is undoubtedly responsible for some sudden, and disastrous, glacier advances. An extreme suggestion is worth notice here, even though it concerns neither valley-glaciers nor temperate glaciers. Portions of the Antarctic ice-cap appear to be underlain by meltwater – melted, perhaps, by the slow escape of the earth's heat beneath the insulating ice. Some observers perceive an imminent danger of massive slides.

Almost all the movement of cold glaciers, and most of the movement of many temperate glaciers, takes the form of internal deformation. In total, such movement passes under the general name of *creep*. It can conceivably include any one, or any combination, of plastic deformation of ice crystals, recrystallization without melting, slide of crystals against one another, re-freezing, and large-scale shear. This last process is easy to document. Sorting out the remaining processes from one another presents difficulties. Which processes operate, and the balance struck among them, depends at least in part on local conditions.

Where a glacier flows over and round a small obstacle, such as is illustrated by the rock-knob in Plate 57, it is thought to accommodate itself by melting and re-freezing. Passage across a large obstacle involves deformation, not necessarily excluding shear. Local conditions often induce thinning or thickening of the glacier. Thinning is termed *extension*, the associated flow conditions being those of *extension flow*. Extreme cases occur where a sudden steepening of down-valley slope is enough to produce an ice-fall. Steeply plunging shear-planes open in the ice at the fall. The opposite situation, *compression*, applies where for any reason the ice is locally thickened: it is associated with *compression flow*. Compression occurs within the ice-portals which drain the Greenland cap through its mountain rim (Fig. 69). Compression also typifies many glacier snouts. As will be seen shortly, it is there accompanied by large-scale shearing, but with shear planes gently inclined.

A striking characteristic common to many valley-glaciers is the linear arrangement of rock-waste on the surface of the ice (Plate 58). Any fragmentary material transported by ice, or directly deposited by ice, is *moraine*. Part of the total load of moraine is torn off the base and sides of the trough by the rock-shod ice itself, but the ridges of fragments visible in the middle reaches of a valley-glacier have fallen from above. They are composed of scree which, wedged off the high bare slopes by frost, tumbles down the steep slopes on to the edges of the glacier. Here it forms lateral moraine. When two glaciers combine, two of the lateral moraines unite in a line of medial moraine. Ten medial moraines on a trunk glacier record ten confluences. Medial moraines show very clearly that turbulent mixing of ice-streams does not occur. The visible deposits extend downwards, enveloping each contributory tongue (Fig. 68). Only near the snout, where melting tends to destroy

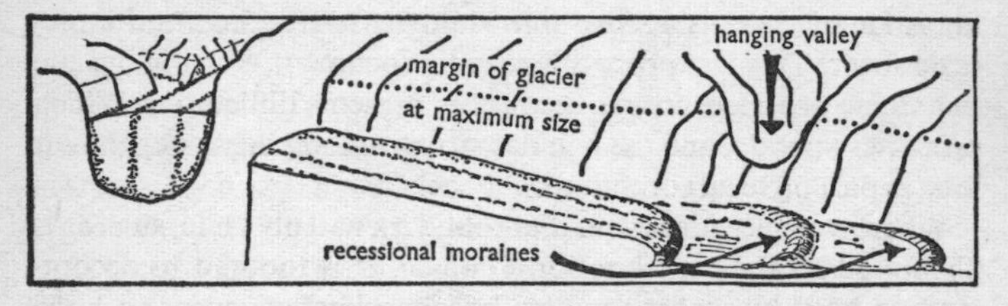


Fig. 68 Moraines of a valley glacier

their identity, may the separate tongues merge in a single mass of very dirty ice.

Unless the ice-front is effectively advancing down-valley, under pressure from excess ice upstream, movement at the glacier end is handicapped. Thinned by melting, the glacier snout cannot easily respond to the pull of gravity. Although some fine-grained debris can be removed by streams of meltwater, the larger fragments remain. As long as the trunk glacier continues to move, rock-waste continues to arrive at the snout, which tends in consequence to become highly charged with broken rock. It yields by shearing, the least mobile basal part being overridden by the cleaner ice behind. But the overriding cleaner ice, in turn, becomes dirty as it is reduced by melting. In this way, deposition at the ice-front is narrowly concentrated in a single band – the band of end-moraine.

Composed of unsorted angular rubble, end-moraines offer no great resistance to erosion by running water. They may, however, be quite bulky – far bulkier than lateral or medial moraines. Thus, whereas end-moraines are numerous in many glacier-troughs, lateral and medial moraines may well be absent or at least unidentifiable. End-moraines alone remain to record glacial deposition and stages in glacial decay. In the fresh state they appear as arcuate embankments. Even where they have been dissected into belts of hummocks, they are easy to recognize. In some troughs they are found at intervals all the way up to the very mouths of corries (Plate 55). Where parallel series of end-moraines can be identified in neighbouring troughs, the sequence of spasmodic deglaciation can be established.

It seems very probable that, in each episode of recession, the extremity of a valley-glacier becomes stagnant. If the rate of supply diminishes, or if the rate of melting increases, the point where supply and loss counterbalance one another is displaced up-valley. At this point a new snout forms. Between the new and the old snouts the glacier, deprived of nourishment, wastes away. The coarser part of its load is left where it melts out, lining the lower parts of the trough with unsorted waste. Deposition directly from the ice is not, therefore, confined solely to the temporary positions of the snout.

When valley-glaciers disappear entirely, and the work of normal erosion is resumed, features of glacial erosion and of glacial deposition are attacked by surface-wash and by post-glacial streams. In time, presumably, the features distinctive of highland glaciation will be entirely re-modelled, and the signs of ice-action wholly destroyed. But at this moment of geological time, many highland areas which now carry no ice-streams have obviously been glaciated in the geologically recent past. They combine glacial features – in varying states of preservation – with a distinctive range of post-glacial forms.

In addition to corries, troughs, lake-basins, rock-steps, rock-knobs, peaks, and ridges, the full range of erosional features includes hanging valleys and truncated spurs. Hanging valleys are the former troughs of tributary glaciers (Plate 62). Because their ice-streams were smaller than the trunk glaciers, the tributary troughs fail to descend to the floor-level of the main troughs. Truncated spurs result from abrasion at the edge of a powerful ice-tongue, which removed large projections in carving its necessary channel. In extreme cases, not merely a spur but the whole side of a mountain has been shorn away (p. 179).

On the sheer hillsides and valley-walls produced by glacial erosion, and on the irregular floors of glacier-troughs, a rather restricted but highly distinctive range of post-glacial features may be observed. Frost-shattering, still operating today, supplies the scree which lies banked against the sides of troughs and the head-walls of corries. Landslides and mud-flows bring down the ice-laid waste lodged on steep slopes (p. 12). Streams discharged by hanging valleys cascade into the main troughs, sawing through the rock at the top of the falls and building fans of debris at the bottom. Shorter streams, rising within the troughs and cutting gullies in their walls, also construct fans which spread across the valley-floors. Deltas form where streams flow into lakes (Plate 63).

Depositional features concentrated at the foot of steep slopes combine with erosion at higher levels to reduce the cross-valley gradients. In many troughs, the visible cross-profile extends across rock in place at the top and across rock-waste temporarily deposited at the bottom. The U-section described as characteristic of glacier-troughs is frequently, therefore, a composite feature representing a partial readjustment of slopes to post-glacial conditions. In any event, very many troughs are only U-shaped when they are seen in oblique perspective. Commonly, their floors are flat in cross-section – the result of the redistribution of rock-waste by streams of meltwater. Meltwater streams issuing from present glaciers usually flow for some distance in braided channels, reworking across the whole width of the floor the material which they transport, and levelling-off its top. Little is known about the depth to which troughs in general are filled with rock-waste. Information from specific sites, which have been explored in connection with projects for damming, shows that some fills are hundreds of feet thick and that the rock-floors beneath are hummocky and irregular.

Some highland regions which have undoubtedly been subjected to severe glaciation prove greatly disappointing on first sight. They fail to include some of those landscape-features which are generally accepted as typical of glaciated highland. One of three causes may be responsible for their morphological deficiencies. Rather rarely most of the signs of glaciation have been destroyed by post-glacial erosion. Occasionally, glacial erosion has operated in a manner different from the manner observed in the Alps. Quite frequently, the underlying rocks have not responded well to glaciation.

The terrain most favourable to the full development of glacial land-forms is that of a highland region, based on strong rocks, where isolated summits and high ridges rise above the snowfields at glacial maximum. In these circumstances, the heads of pre-glacial valleys can be converted into corries. But not all glacier-troughs lead upwards into corries. Some become progressively shallower up-valley, merging finally into the general surface of some high plateau. Such troughs are the work of tongues of ice draining from flat-lying local caps, in country where the pre-glacial summits were very broad and very subdued.

In the Pennines and in central Wales, the main pre-glacial valleys were deepened, and to some extent widened, by ice-tongues. At glacial maximum, when ice was deep enough to cover most of the watersheds as well as to fill the valleys, the conditions were those of sheet-glaciation rather than those of sculpture in Alpine conditions. But while this fact doubtless goes far to explain the lack of corries, it cannot account for the absence of corries from the flanks of high mountains which rise above the plateaus. Ingleborough, for example, is certainly high enough to have been exposed to corrie-cutting, but the shaly rocks of its upper part seem to have been incapable of developing – or of preserving – the sheer faces of corrie headwalls. Furthermore, although the Pennine and central Welsh valleys are recognizable as glacial grooves once they are known to have been glaciated, many of them do not look, in the field, particularly impressive as evidence of sculpture by ice. Dark boulders of grit, deposited by melting ice on the pale-grey, well-weathered limestone of the lower flanks of Ingleborough, seem incongruous (Plate 64). Hundreds of marginal spillways on the flanks of large valleys appear equally out of place. Were it not for such features, the subdued signs of ice-erosion might be overlooked altogether. Neither the Carboniferous rocks of the Pennines nor the generally fine-grained, shaly rocks of mid-Wales are capable of recording glaciated highland at its best.

## CHAPTER THIRTEEN

# *Ice over All*

ICE-CAPS can be 2 miles thick. This figure, calculated from the physical properties of ice, may have to be modified when the results of work in Antarctica become known. But it certainly seems to be of the right order, judging by recent observations on the existing cap of Greenland.

Soundings give a maximum thickness for the Greenland ice of about 11,000 feet (Fig. 69). Seismic methods of sounding are those

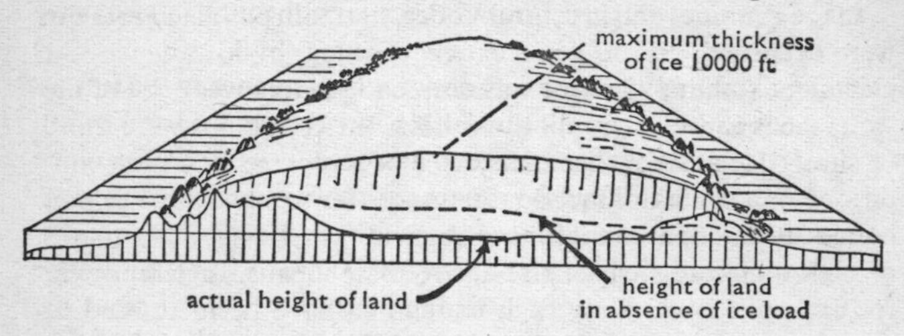


Fig. 69 Depression beneath the Greenland ice

chiefly employed. Shock-waves are generated near the surface of the ice by the detonation of small explosive charges. The time taken for the waves to travel down through the ice, and to be reflected back to the surface from the underlying rock, is electrically recorded. The technique is essentially similar to that used to record earthquake-waves, and the explosions are, in fact, small private earthquakes.

In addition to giving a maximum thickness for the ice of some 10,000 feet, the soundings reveal the form of the floor on which the ice-cap rests. The Greenland cap is contained in a great shallow basin, with raised edges appearing at the margins of the cap as mountains (Fig. 69). Some of the mountains exceed 10,000 feet in height, whereas the land beneath the centre of the cap lies, in places, below sea-level. This arrangement agrees perfectly with what is known of ice-caps which have melted away. Under their great weight the crust of the earth was depressed.

A layer of ice 10,000 feet thick weighs as much as a layer of average rock about 3500 feet thick. By means of measurements of gravity, it is possible to show that if the ice were no longer present the rock-floor beneath the Greenland ice would stand 3500 feet higher than it now does. Subsidence compensates fully for the ice-load, which in Greenland amounts to some 650,000 cubic miles, weighing some 2,500,000,000,000,000 tons.

But subsidence does not immediately follow the formation of an ice-cap. A minimum load must be applied before the crust will sag. Subsidence beneath a growing cap is therefore intermittent. Conversely, when an ice-cap disappears, recovery follows; but it is both intermittent and delayed, so that the rise of the land is tardy and spasmodic.

It is proposed to illustrate this matter by reference to the former Scandinavian ice-cap, about which a very great deal is known. Indeed, the late history of that cap has been so well reconstructed that the recent work on the existing cap of Greenland fully confirms the principles established by Scandinavian researchers.

At the glacial maximum of about 200,000 years ago, ice covered the Arctic Ocean, merging with the land-ice of North America, Greenland, northwest Europe, and northern Russia (Fig. 70*a*). The

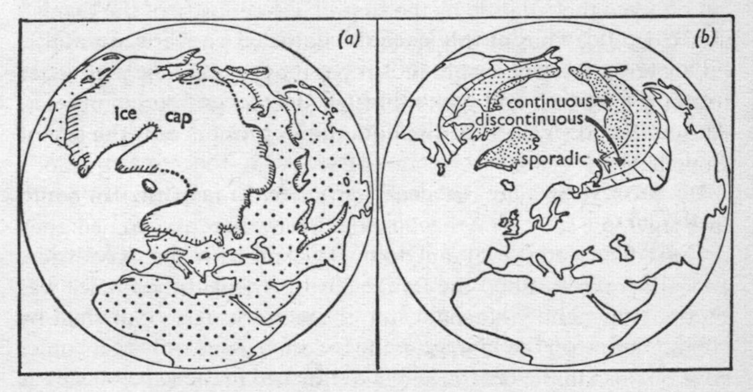


Fig. 70 (*a*) Ice-caps at glacial maximum
(*b*) Permafrost today

North American sheets, spreading from more than one centre of dispersal, reached far southward across the continent. The whole of Greenland was overwhelmed. A separate cap concealed Iceland. Norway, Sweden, Finland, Denmark, and the Baltic Sea were blanketed by an ice-sheet which sent huge lobes towards the south and united across the North Sea with the ice which covered most of the British Isles.

This was the last but one of the four glacial maxima of the Ice Age. It was followed by an interglacial period of warmer climate, which was in turn succeeded by the last great re-advance of sheet-ice. The last glacial maximum failed to re-create ice-caps on the former extensive scale, although one large sheet formed over the Baltic and its borderlands, and a separate cap was established over the northern parts of the British area.

So much water was locked up in the form of ice that the level of the sea throughout the world was lowered. At this last maximum, the sea stood 200 or 300 feet lower than it does now. The English Channel and the southern floor of the North Sea were dry land, subjected to a very severe climate and with their subsoil permanently frozen.

By about 8000 B.C. – as little as 10,000 years ago – the melting Scandinavian ice had uncovered the southern Baltic. Because sea-level was still below the present mark, the shallow Baltic basin was cut off from the Atlantic by the chaotic ice-deposits of the Danish area (Fig. 71). Thus at this stage the Baltic was a freshwater lake, little above freezing-point in temperature, receiving meltwater from the ice-front, and with icefloes and icebergs floating upon it. At first the lake spilled across hummocky ground near the site of Copenhagen.

By 9700 years ago the decaying ice had uncovered in south Sweden a low sill, through which the freshwater lake immediately drained. The water-level sank to the level of the rising sea outside, and salt-water invaded the Baltic Basin. By this time the ice was becoming generally stagnant, for it was no longer nourished by copious snow on the high ground. It decayed not only at its outer edge but also in the centre, and lakes formed in the valley-heads as it melted off the main watershed.

Another 1000 years saw the ice almost gone. By 8500 years ago the lakes in the mountain valleys had been released. Post-glacial time is considered to have begun when the ice-cap split into two, and the Jämtland lake flowed through the breach into the Baltic

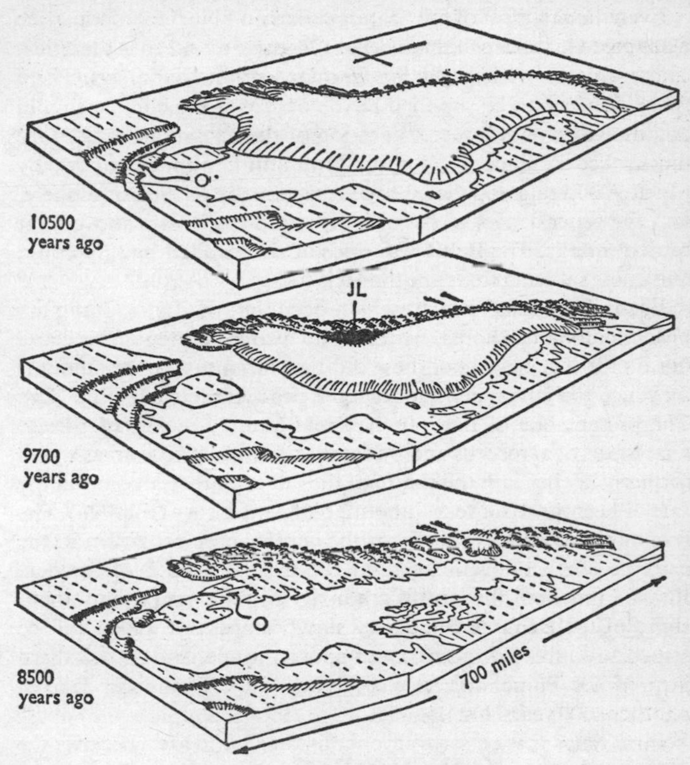


Fig. 71 Deglaciation of Scandinavia

Basin. Meanwhile, the general level of the sea was being raised by the inflow of meltwater from all the glaciers of the world. If this had been the whole story, the entrance to the Baltic would have been widely flooded. But the Scandinavian land, relieved of the weight of ice, was recovering. The broad but shallow channels leading to the west were first narrowed and then dried out, so that the Baltic once more became a freshwater lake. In the extreme south it was enclosed by a narrow neck of land uniting Sweden with Denmark, and its outlet ran through the south Swedish lakes and down the valley of the river Göta.

Even though most of the Scandinavian and North American ice disappeared, ice-caps remained in Greenland and in Antarctica. There were also bulky glaciers in many mountain valleys. There was still water to be supplied to the oceans if the climate should continue to grow warmer. Warming of the air and melting of ice did in fact continue, so that sea-level still continued to rise. By about 7000 years ago the average temperature of the atmosphere, and the general level of the oceans, stood somewhat above their present marks. The Baltic was once again invaded by salt-water, becoming saltier at that time than it is today.

Each of the seas or lakes which occupied the Baltic Basin cut beaches along its shores, wherever the water abutted on land and not on ice. The heights of these old beaches measure the amounts by which the land has risen since the beaches were cut (Fig. 72*a*). The present rate of rise can be measured by means of precise analysis of tidal records and by detailed survey. Those areas which formerly lay beneath the centre of the ice-cap are recovering at the rate of 1 centimetre a year – about 3 feet per century (Fig. 72*b*). The rate of rise declines away from the centre, becoming zero in the extreme north of Denmark and the extreme south of Sweden. Beyond this limit the land is gradually sinking, and the southern shore of the Baltic is being very slowly flooded. Sweden will be joined to Finland if these movements continue, and the northern arm of the Baltic will once more become a freshwater lake in another 8000 years' time.

Firm dates can be set to events in the Baltic area because the decay of the Scandinavian ice-sheet year by year has been measured. Two principal methods of counting are available, each depending on a special kind of deposit. In the last stages of decay an ice-sheet is stagnant, melting away where it stands. But while it is still receiving increments of snow in the central areas the ice still tends to move outward, even though the outward movement may be more than counterbalanced by melting (Plate 76). Melting-back is concentrated in the summer season. During the late winter and early spring, the ice-front can remain stationary or even advance a little. Consequently, each annual position may be marked by a line of debris melted out of the ice – an annual moraine, or even by a small ridge ruckled up during the slight advance – a push-moraine (compare Plate 65).

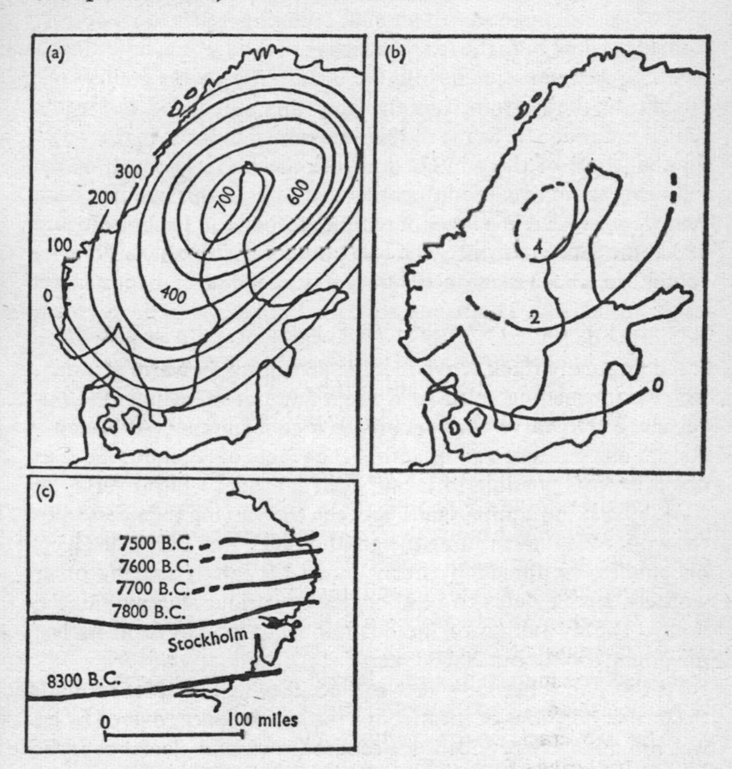


Fig. 72 (*a*) Total elevation of Scandinavia, in feet, during the last 9700 years
(*b*) Present rate of rise, inches per 10 years
(*c*) Dates of moraines near Stockholm

If minor moraines of these kinds could be traced widely across country, the history of glacial decay would be simple to establish. But as the morainic features are discontinuous and scattered, the year-by-year displacement of the ice-front cannot be followed in full detail. On the other hand, far more bulky moraines occur where the front remained more or less stationary for some length of time, so that the general progress of deglaciation is well recorded (Fig. 72*c*.)

The second method of annual counting relies on deposits formed in lakes, or in the sea, close to the ice-edge. Each summer's melting released sediment into the water. The coarse grains sank directly to the bottom, but the very fine suspended fragments settled more slowly. Some of them remained diffused in the water for the whole of the winter, until the increasing warmth of the following summer caused them, too, to sink. In this manner, each year was recorded in a layer of sediment, coarse at the bottom and fine at the top. Frequently the two parts of a single layer differ in colour, so that a section through a succession of bands looks striped (Plate 66). The bands are called *varves*, from the Swedish *varvig*, banded.

An unusually thick varve records an unusually warm summer, when extra melting released a large amount of sediment. Conversely, a thin varve corresponds to a cool summer. Although a continuous series of varves across the whole of Scandinavia does not exist, it is possible to identify partial series which overlap in time, by making comparisons between the varying thicknesses. In the work of varve-counting the great pioneer was Baron de Geer; his studies, continued by members of his family and by other workers, enable dates to be allocated to various stages of glacial decay. The dwindling of the Scandinavian ice-cap from its last maximum took about 20,000 years.

As this and other ice-sheets melted, leaving moraines to mark the former positions of their fronts, the ground once covered by ice was progressively revealed. Ice-scoured rock and glacial deposits came into view. In addition to the frontal moraines, there were large accumulations of sand and gravel washed out from the melting ice, in the form of extensive flat sheets or disorderly mounds.

One particularly remarkable class of deposits consists in the sandy and gravelly ridges called *eskers*. A distinctive name is required, for eskers occur by the hundred in Scandinavia (Plate 68). In a rough sort of way, the Swedish and Finnish eskers are aligned at right angles to the former ice-fronts. They correspond to the channels of streams which ran over the surface of the waning caps, or along tunnels inside or beneath the ice. As long as the ice-sheets lasted, the sandy and gravelly linings of the stream-beds were supported by frozen walls. But when the ice decayed, the well-washed deposits were left as ridges rising above the level of their immediate surroundings. Where a stream debouched at the ice-front, its charge of sand and gravel spread out fanwise. Bulky fans were built on lines where the ice-front remained stationary for a time, so that eskers swell out where these lines are crossed.

The Swedish name for a ridge of this type is ås – plural, åsar. English geologists, rendering the words *os* and *osar*, chose first to make *osar* singular and then to equate it with the English word *esker*. In this manner, the word esker has become specialized to mean a ridge of sand and gravel deposited at right angles to the ice-front. But the word esker arrived from Ireland, being derived from the Gaelic *eiscir*. This latter term was originally applied to the great numbers of gravel ridges which curve, arabesque-like, across the Central Plain of Ireland. These ridges too are of glacial origin. They were however laid down along the ice-fronts and not at right angles to them. The name used for accumulations of sand and gravel deposited against the ice-front whether as individual mounds or as long narrow belts is the Scots word *kame*. By common consent a distinction is made between kames and eskers. The fact that most Irish eskers are really kames – in the accepted technical meanings of these words – cannot be helped.

Stagnant ice-sheets decayed in an irregular manner. Tension-cracks in their spreading lobes were widening by melting, and the sheets disintegrated into separate blocks. In some areas, outwash deposits choked the widening fissures in decaying ice, and completely buried the remaining blocks under flat sheets of sand and gravel. When, in due course, the concealed blocks also melted, the overlying sand and gravel slowly collapsed to form hollows. This process can be seen at work today on the lobate foot of the Malaspina Glacier of Alaska.

Plains of outwash can become pitted by multitudes of depressions, which contain lakes wherever the level of the water-table is high enough. One of the finest examples of a pitted outwash plain is to be seen in the neighbourhood of Minneapolis (Fig. 73).

Hundreds of lakes occur in this area, most of them small but a few quite sizeable. The waves have carved sandy beaches at the edge of the enclosing deposits, which between the lakes are hummocky and clothed with softwood trees. Round blue lakes, bright circular beaches, and solid dark masses of trees combine in an attractive and memorable scene.

Within the limits of the outer moraines, the wholesale melting of stagnant ice allowed the loads of rock-waste to settle gradually on to the underlying surface. The general result is a spread of completely unsorted material, ranging in size from minute clay particles to enormous blocks of rock. These are the constituents of

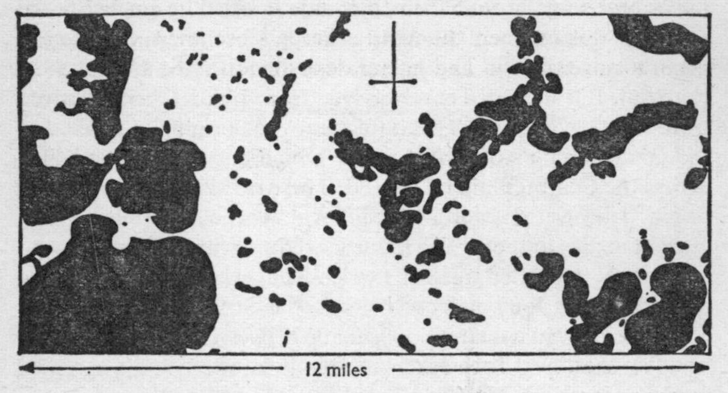


**Fig. 73 Kettles near Minneapolis**

boulder clay (Plate 60). Sheets of boulder clay in East Anglia, which lay at times beneath the margin of the Scandinavian sheet, attain total thicknesses of 300 feet. This figure represents the sum effect of all the glacial advances of the last ice age. Fossil boulder clay in Australia, deposited during an earlier ice age dating from 250 million years ago, reaches thicknesses of no less than 100 feet.

Boulder clay varies according to the nature of the ground over which the ice formerly moved: some boulder clays are extremely stony, others consist mainly of clay which has been re-worked by the ice. In any event, any given boulder clay is certain to contain glacial erratics – rock fragments carried by ice far from their original sites.

Local material is normally the most plentiful. The boulder clays in the north of the London Basin provide abundant specimens of the local Reading Beds, in addition to blocks of Chalk and very numerous flints derived from the Chalk beds. But some of the flints, and occasional pieces of Chalk, belong to the red basal Chalk of Lincolnshire. Sandstone erratics have been introduced from beyond the Chilterns, and various limestone erratics include blocks transported from the limestone hills of Lincolnshire and Northamptonshire. Pebbles of quartzite and vein-quartz recognizably belong to a series of pebble beds which occurs in the Midlands; their most probably source is Sherwood Forest.

Certain erratics enable the net direction of transport to be accurately determined (Fig. 74). Among these are pieces of rock from Mountsorrel in Leicestershire – the so-called Mountsorrel

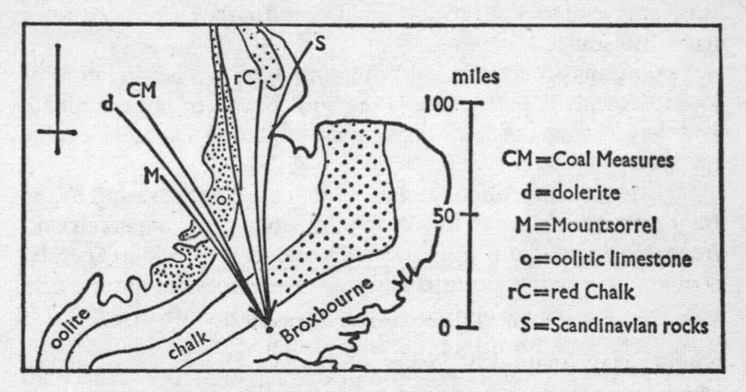


Fig. 74 Sources of some Broxbourne erratics

Granite, which occurs only at this one spot. Similarly, limestone from Derbyshire can be certainly identified. The Derbyshire Dome in the south Pennines is the most likely supplier of blocks of dolerite. All the evidence so far mentioned points to glacial transport from the north.

Very far-travelled erratics include blocks from the east of Scotland – the eastern part of the Scottish Highlands or the Cheviot district. There also record movement from north to south. But the southward-streaming ice on the eastern side of Britain was guided by the Scandinavian ice which spread across the North Sea Basin. Its former presence is abundantly recorded by erratics of Scandinavian rock, among which in the London Basin occur fragments of rhomb-porphyry. This distinctive Norwegian material, pinkish or purplish in colour, contains large angular crystals of the mineral felspar.

Identifiable erratics show the net effects of ice-transport, but cannot indicate the precise lines followed by the ice in any one advance. Since there were repeated advances, some of the debris has been shifted more than once. Although each advancing sheet roughly followed the directions previously taken on earlier occasions, it cannot safely be assumed that erratics were carried in straight lines from where they were torn away to where they were finally dropped. Distributional studies prove that blocks from a small area tend to occur in a widening band as they are traced away from their source of origin.

Two means are available to determine the precise direction of ice-movement. The structure of the boulder clay can be examined, and the scratches made by moving ice on underlying rocks can be mapped.

Strictly speaking, the scratches are made not by the ice but by the fragments which it carries. A glacier, shod with angular rock-fragments, acts like a huge rasp. In addition to being generally scraped by moving ice and abraded by sharp sand-grains, firm rock beneath the ice can be deeply scratched (Plate 61). But it is possible to work for long periods in glaciated areas without seeing a single scratch. Except in regions of very resistant rock, and within the limits of the last great ice-caps, weathering has commonly destroyed any glacial scratches which may once have been present. Some rocks are so highly jointed that it is impossible to distinguish scratches from fissures. Even where they occur, scratches can only indicate ice-movement in one of two directions, and since basal ice can move locally uphill, evidence other than the arrangement of scratches is needed in certain critical circumstances (Chap. 14). Nevertheless, when glacial scratches are mapped throughout a large area, as they have been mapped in Sweden during a period of 100 years, a plot of their distribution can be used to show, with some accuracy, the direction of ice-movement.

The second means of establishing direction of movement is most likely to be effective precisely where scratches on solid rocks are absent – that is to say, in those areas where thick formations of weak rocks, most unlikely to display scratches, are in any case concealed by a thick sheet of boulder clay. It has been known for more than a century that stones in boulder clay typically lie with their long axes parallel to ice-scratches on the stones themselves. Thus the arrangement of the boulders indicates the direction taken by the moving ice.

This principle was first effectively applied in north Germany in the 1930s, during an investigation of the great end-moraines of Pomerania. Subsequently it was used by researchers in North America, Sweden, and Iceland. Every application was successful. More recently, it has been used to subdivide and correlate the boulder-clay sheets of East Anglia and the east Midlands. Shallow pits are dug in undisturbed boulder clay, the bottoms being carefully cleaned. The arrangement of boulders is measured by compass, and a diagram constructed for each site. It is almost invariably found that the long-axes of the boulders point in a single dominant direction. When the results for numbers of pits are plotted on a map, a picture is obtained of the pattern of ice-movement. This type of analysis can be of the greatest use in separating from one another two boulder clays present in a single area.

Where the boulder clay has been moulded by overriding ice, in a rhythmic fashion which cannot yet be explained, a well-defined type of landscape is produced. Oval hills are formed, tapering away towards their narrow ends, which point in the direction of ice-movement. Hills of this kind are drumlins. Since the word *drumlin* is Irish, the typical occurrence of drumlins may suitably be illustrated from Ireland (Fig. 75*a*, Plate 69). Drumlins often occur in groups, or swarms, any one of which may contain hundreds of drumlins.

Hollows in a drumlin-swarm are likely to be ill-drained, containing bogs or even lakes. In the basin of the Erne a drumlin-swarm is partly flooded, so that the hills rise from the interlacing waters of a shallow lake of most unusual shape (Fig. 75*b*).

In a general way we have been working steadily inwards, from the front of ice-caps towards the central areas. It remains to say what happens beneath the middle of the cap. At glacial maximum it is possible for the central ice to be almost stationary. Snowfall may be concentrated on the margins of a great cap, which so severely chills the overlying air that travelling depressions cannot invade it. But during its growth, and perhaps also during its decay, the ice in the centre is likely to move outwards, heavily scouring the underlying surface.

In detail, the scouring action is highly selective. Weak rocks and weak structures are eroded, while resistant rocks withstand the attack with more success. Consequently, the terrain revealed by

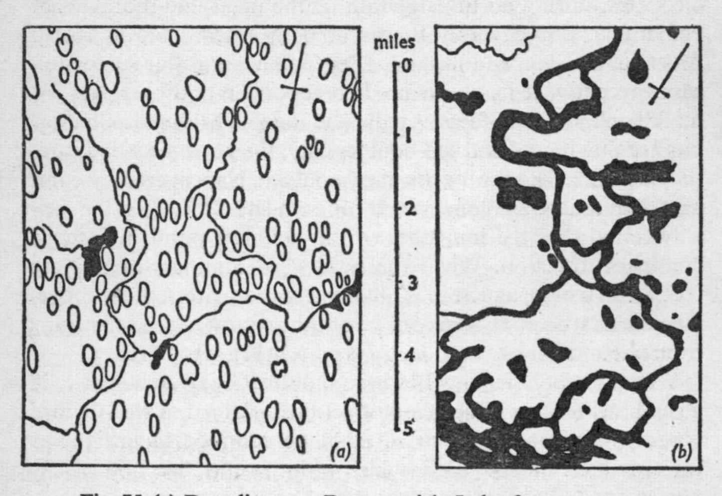


Fig. 75 (*a*) Drumlins near Downpatrick, Ireland
(*b*) Partially drowned drumlins near Enniskillen, Ireland

deglaciation is broken by minor irregularities, as is shown very clearly indeed in Finland and in the Canadian north. The regions in question are based on greatly altered and generally tough rocks, but are traversed by well-developed systems of faults and joints. Even where the post-glacial drainage is free from the obstruction of glacial deposits, it is characteristically disordered. Far too short a period has elapsed since the disappearance of the ice for the drainage to have integrated itself. Bare rock is widely exposed, the pre-glacial waste-mantle having gone to form the deposits of the outer and marginal districts. Stream-courses are broken by innumerable falls. Rivers combine in an intricate net, lakes and swamps occupy the enclosed hollows, and water covers a large fraction of the surface.

*

At various points in the foregoing section, reference has been made to the occurrence of more than one glacial maximum. The Ice Age involved more than one episode of glacial growth and glacial decay, as has long been evident from the relations of glacial deposits. Not only are successive advances recorded by separate sheets of boulder clay – long periods of quite genial climate are indicated by deposits occurring between successive sheets. Again, the older sheets are known to have been deeply weathered in the interval between their deposition and the next glacial advance. Hence the Ice Age is subdivided into a number of glacials and interglacials. During interglacial times, the climate was at least as genial as it is today.

When attempts are made to draw up a scheme of successive glacials and interglacials, serious difficulties immediately arise. The original scheme was worked out for the Alpine glaciers, which are generally recognized to have undergone four episodes of great extension. Thus the Alpine sequence is subdivided into (1) preglacial, (2) first glacial maximum, (3) first interglacial, (4) second glacial, (5) second interglacial, (6) third glacial, (7) third interglacial, (8) fourth glacial, (9) post-glacial. But because the Alpine glaciers never made contact with Scandinavian ice, no glacial deposits can be traced all the way from the Alps to Scandinavia, and correlations between the two areas have to be made by indirect means.

It is frequently difficult to decide which deposits represent a glacial maximum, and which correspond to lesser re-advances. The freshest deposits, left by ice at the last maximum, show quite clearly that both the expansion and the decay of the Scandinavian sheet were irregular. Similar evidence is forthcoming from the Alpine area. Events associated with earlier maxima were presumably no less complex. Some early unsuccessful attempts to relate developments in northern Europe to those in the Alps showed that, for the time being at least, an independent scheme must be drawn up for the Baltic glacials, and local names were given to the stages recognized there. A third set of stage-names was established for the Netherlands. In Great Britain matters were greatly confused by the simultaneous use of Alpine, German, Dutch, and local British terms. Yet another scheme was evolved in North America, where – as in the Alps – four glacials were identified; these were named the Nebraskan, Kansan, Illinoian, and Wisconsin glacials, after states where their deposits are particularly well preserved and exposed.

So long as the glacial deposits of north and northwest Europe were divided into four groups, it seemed easy to equate them with the four groups left during the four extensions of glaciers in the Alps. But accumulating evidence made it impossible to distinguish more than three glacial maxima in northern Europe, and it is now held that the first of the possible four glacials has left little or no trace in the Baltic region. If a Scandinavian ice-sheet developed at that early stage, it seems to have been smaller than later sheets, its deposits being destroyed by weathering or wholly re-worked by later ice.

For a long time it remained uncertain that the ice-caps around the North Pole had waxed and waned simultaneously with those around the South Pole. It is now generally held that glacial maxima occur simultaneously in the north and the south hemispheres. Similarly, it is held that glaciation was simultaneous in Europe and in North America. Even so, there is no agreement on the absolute dates of the varous glacial maxima. Within the last 25 years or so, eminent authorities have placed the beginning of the first North American glacial as early as 1 million years ago and as late as 500,000 years ago. However, there seems little doubt that the second of the three interglacials in North America was a long one. The second interglacial is known to have been lengthy in the Alps, where it is called the Great Interglacial. This being so, it seems likely that a generally agreed scheme will eventually be adopted for the history of glaciation on both sides of the North Atlantic.

In all probability, such a scheme will resemble the summary made in Fig. 76. Dates given in this diagram are due to Zeuner, who is the main advocate of explaining the alternation of glacials and interglacials by means of changes in solar radiation. Two points should be carefully noted. The changes referred to are changes in radiation received by the earth, not changes in solar emission. While they can account for the differences between glacial and interglacial conditions, they cannot explain why there should be an Ice Age in the first place.

Three astronomical causes act to vary the amount of solar radiation received at the outer limit of the earth's atmosphere. They are periodic, but have periods of different length, so that sometimes they act in combination and sometimes in opposition. A slight variation in the angle between the earth's axis and the plane of its orbit affects the character of the seasons. The second

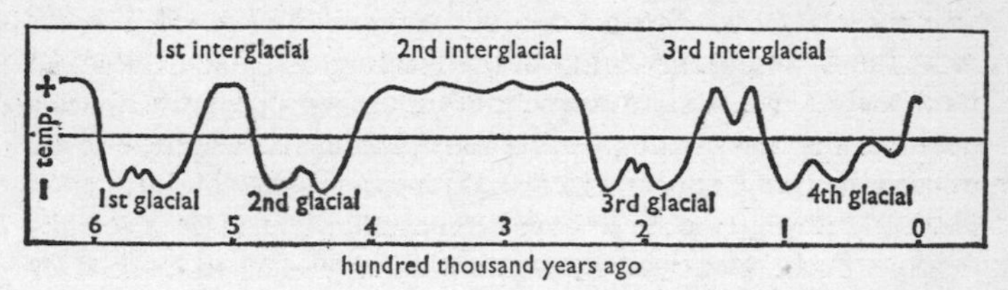


**Fig. 76 Variations of temperature during the Ice Age (adapted from Zeuner)**

variation affects the shape of the orbit, which at times becomes more nearly circular and at others more elliptical. The third factor is a slight, long-term roll of the spinning earth, which alters the orbital position of solstices and equinoxes. Since the respective periods of the three kinds of movement are 40,000 years, 92,000 years, and 21,000 years, their resultant effect can range widely.

Not everyone agrees with Zeuner's view that this effect serves to explain and to date glacial fluctuations. Among the dissentients is Öpik, who relies not on variations in orbital elements but on changes in solar luminosity. Taking the question still farther back, this writer concludes that major changes in luminosity repeat themselves at intervals of about 250 million years – that is, with a periodicity capable, it would seem, of explaining the buried record of ice ages far earlier than the one in which we live.

Much can be hoped from methods of radiometric dating that can do for the last one or two million years what radiocarbon dating is doing for the last seventy-two thousand years. Among the methods applicable to part of the desired time-span is the isotopic analysis of thorium and uranium ($^{230}Th/^{234}U$). For the Barbados reef tracts, this method during the late 1960s produced periods of coral growth that closely match the calculated maxima of summer solar radiation, about as far back as a quarter of a million years. Accordingly, the hypothesis of astronomical control is, to say the least, powerfully revived.

CHAPTER FOURTEEN

# *Glacial Interference*

GLACIERS in mountain valleys, and ice-sheets spreading over the surface of the land, occupy country which in pre-glacial times was being modelled by running water. Drainage is inevitably disorganized. At the very least, rivers are temporarily replaced by tongues of ice. The land-surface is considerably modified in detail, and in places undergoes drastic alteration – alteration which affects the courses of post-glacial streams. At the other extreme, whole new landscapes are formed by glacial deposits; the drainage-systems developed on them when the ice melts have little resemblance to the pre-glacial patterns.

It is appropriate here to re-emphasize the episodic character of glaciation. Valley-glaciers and ice-sheets obtrude themselves on the landscapes of normal erosion. They endure for periods which, by comparison with the span of an erosion-cycle, are very short. In consequence, they interfere with the normal pattern of drainage, and modify the normal landscape. On deglaciation, running water renews its work. It can be expected that, in due course, the marks of ice-erosion will be destroyed. But it so happens that we are living at a time when the signs of ice-action are still fresh. Among those signs are many diversions of drainage.

Valley-glaciers descending well below the snowline obstruct the flow of streams entering from the side. Lakes are often formed in the ice-blocked side-valleys (Fig. 77, Plate 58). Such lakes may find outlets along the margin of the ice, but not infrequently they spill across low points in the bordering hills.

One well-known case is that of lakes in the Scottish glens north of Ben Nevis. Vigorous glaciers, descending from the great corries on Ben Nevis and neighbouring mountains, invaded and blocked the lower ends of ice-free glens. Lakes confined by the ice-dams fell spasmodically in level, as the decaying ice revealed lower and lower outlets. Old lake-shores are marked by narrow lines of beach, which run along the mountainsides towards the spillways.

Where long valley-glaciers underwent intermittent decay, spillways leading from one side-valley to another can be used to reconstruct the history of deglaciation. Detailed reconstructions have been made by Raistrick for glaciers in the Pennines, where conditions greatly favoured the formation of temporary lakes. But although some sets of evidence for marginal channels are still accepted, other sets are being re-interpreted as indicating subglacial channels, both for valley glaciers and for ice-caps. Reconstructions of the course of ice-decay are being revised accordingly.

Drainage in glaciated highland can be diverted by ice-erosion as

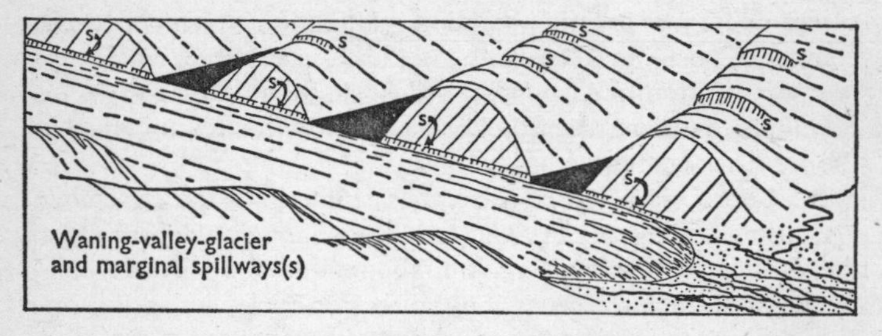


**Fig. 77 Valley-glacier and spillways**

well as by ice-obstruction. Furthermore, diversions due to erosion can be permanent. Some involve no more than minor rearrangements of the drainage-pattern, but others relate to complete piercing of pre-glacial watersheds.

Valley-glaciers might seem unlikely to make permanent changes in the network of highland streams. Guided along the lines of preglacial valleys, tongues of ice move from high ground to low. Trunk glaciers occupy the largest of the pre-glacial valleys, and tributary glaciers emerge from valleys which, in pre-glacial times, carried tributary rivers. In a general way, the pattern of post-glacial streams is likely to resemble the pre-glacial pattern, however greatly the valleys have been modified by ice-action.

But highland ice was not everywhere successfully confined by the walls of pre-existing valleys. In certain conditions it was able to cross watersheds, and even to cut great breaches in them through which post-glacial rivers flow (Plates 70, 71). Glacial breaches in

Britain have attracted much notice in the last 25 years or so. Dozens of them have been identified, although no more than a few have yet been described in print. Obscurities in the patterns of highland drainage have suddenly been made clear.

A particularly clear illustration of the problems set by glacial

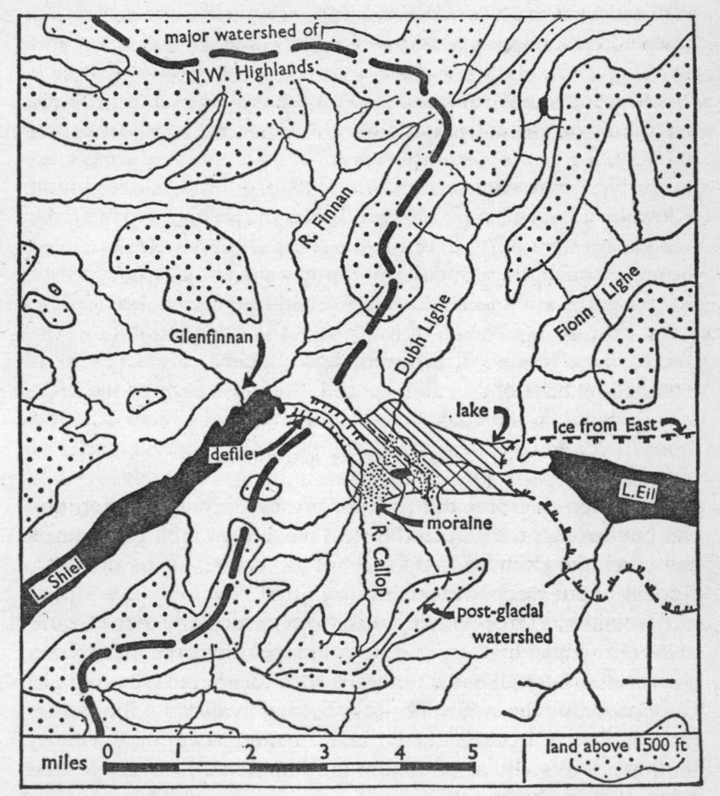


Fig. 78 A stage in the diversion of the Callop (re-drawn after Dury)

breaches is given by the river Callop (Fig. 78). This rather small stream rises in the Scottish Highlands west of Fort William, flowing down the eastern side of a line of hills into the valley which contains Loch Eil. But instead of uniting with the Dubh Lighe – a stream coming from the far side of the valley – and flowing into the lake, the Callop turns sharply westwards and pierces the high ground in a narrow gap. The high ground, based on particularly resistant rocks, forms part of the main mountain axis of northern Scotland. It is impossible to account for the breach, and the westward course of the lower Callop, by any system of river-capture. As soon as the ground is inspected, however, matters at once become clear.

Numerous rock-knobs, scoured on the east and plucked on the west, show that ice moved *up* the Eil valley – under pressure, presumably, from copious glaciers nourished from the deep corries on Ben Nevis. Overriding the pass at the valley-head, the ice ground it down to a level of some 250 feet. When the ice began to melt, the side-valleys were partly freed while a bulky glacier still occupied the main trough. Impounded water poured over the pass, cutting a slit-like gorge through its floor, and opening a deep trench leading to the west. A re-advance of ice caused the sides and floor of the trench to be scratched, scoured, and plucked. Final retreat is recorded by belts of moraine spaced along the floor of the main trough. Between the Callop and the Dubh Lighe there exists no more than a low barrier, consisting partly of moraine, partly of outwash, and partly of the delta-fan built by the Dubh Lighe. It seems a mere accident that, at the present day, the Callop discharges through the gap to the west while the Dubh Lighe turns eastwards into Loch Eil.

Ideas about glacial breaching were first effectively introduced into Britain by Linton, who applied them particularly to the Grampian Highlands. Although the principles involved have been known elsewhere for some time, the additional examples about to be quoted are mainly British – chosen partly because the studies made in Britain are recent, and partly in order to show how greatly the history of river-development is illuminated when account is taken of glacial breaching.

The Spey has often been cited as a piratical stream, which by successive captures has greatly enlarged its basin at the expense of its less successful neighbours. The supposed captures depend on the interpretation of sharp bends in the river-pattern. But progressive capture only seems likely so long as the structural pattern of the Spey Basin is ignored. Like so much of the Scottish Highlands, this basin is criss-crossed by lines of structural weakness. Streams adjusted to such lines are bound to turn through sharp angles as they pass from one alignment to another. The angles do not represent the results of capture, but merely reflect close adjustment to structure of the underlying rocks.

Linton's re-examination of the ground has provided much new evidence and has led to convincing new interpretations. For instance, the upper Geldie has been diverted across the line of the pre-glacial watershed (Fig. 79). Formerly a tributary of the Dee,

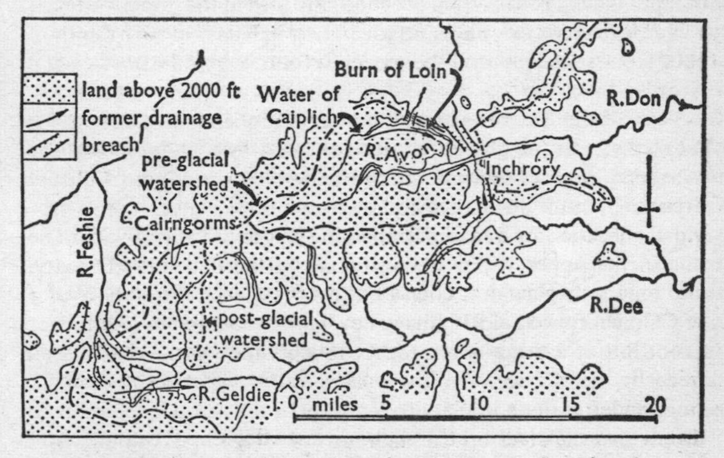


**Fig. 79 Diversions in the Grampians (re-drawn after Linton)**

the diverted river now constitutes the head of the Feshie, a feeder of the Spey. The breach through which the present upper Feshie discharges to the northwest has been opened by glacial erosion.

During glacial maximum, ice coming from the Cairngorms on the north, and from the main central block of the Grampians on the south, accumulated in the valley. Overtopping the pass which, at a level of some 2000 feet, led into the then valley of the Feshie, the ice spilled out as a powerful tongue. A deep trough gouged in the floor of the pass now guides the waters of a diverted stream – the present upper Feshie. Between the Feshie and the deprived Geldie, the watershed is composed of part of an alluvial fan. Here is illustrated a typical feature of glacial breaches, already noted in the case of the Callop – namely, the very minor and accidental nature of the post-glacial watershed on the floor of a glacial breach.

About 20 miles away to the northeast, part of the former Don has been diverted to the Spey. The old head of the Don is represented today by the uppermost 15 miles of the Banffshire Avon. Although the upper Avon is directly aligned on the Don (Fig. 79), the river turns abruptly northwards at Inchrory to flow through a narrow breach in a belt of high ground. The breach marks the site of a pre-glacial pass. When deep ice gathered in the upper valley, it was able to cross this pass and to cut through its floor to a depth of 1000 feet – so deeply that the breach is followed by the post-glacial river.

West of the Inchrory breach lies the Caiplich gap. It contrasts markedly with the glacier-trough followed by the Avon, for it is V-shaped in cross-section. It is not the direct product of glacial breaching, but is due to erosion by overspilling water. It once provided an outlet for a lake dammed in the upper valley by ice. The line of the upper valley is continued by the flat-floored, peaty, and obviously glaciated Glen Loin. In pre-glacial times the Water of Caiplich flowed along this valley as the headstream of a longer Loin. But at a stage when the Caiplich valley had already been freed by melting ice, the lower Loin valley was still obstructed. Impounded water spilled across a pass to the north, eroding it so deeply that the diverted Caiplich has never regained its connection with the Loin.

Glacial breaching, in highland areas, is most likely to occur where deep ice becomes congested in a large head-valley. In the Snowdon district, numbers of head-valleys had been widely excavated in pre-glacial times, and circumstances during glacial maximum were particularly favourable to breaching. Breaching occurred so extensively that the existing drainage-pattern seems, in a rough sort of way, to be radial; but in pre-glacial times it was markedly centripetal, with bold upstanding mountains encircling open valley-heads and enclosing integrated sheaves of headstreams.

Two neighbouring gaps near the western flank of Snowdon respectively illustrate minor and major effects of breaching. The first gap connects the valley of the Gwyrfai, at the head of Llyn Cwellyn, with the glacier-trough drained by the Llyfni; the second lies at the lower end of Llyn Cwellyn, running northwestwards towards Bettws-Garmon (Fig. 80). In pre-glacial times these two gaps were passes over the watershed which bounded the Colwyn Basin. At glacial maximum, ice descending from the surrounding mountains deeply filled the upper Colwyn valley, overflowing and eroding both the passes. That leading to the west – the Drws-y-Coed col – was deeply gouged. Its sides were shorn back, and knobs of resistant rock projecting from its floor were severely scoured on the eastern side and severely plucked on the west. Within the existing pass the watershed is ill defined, with two small lakes enclosed in ice-eroded basins. If a corrie was ever cut at the head of the Llyfni valley, its forms have been largely destroyed by the invading ice which crossed the watershed.

Glacial scour in the second gap was very strong. At the sides of the gap, tough rock offered great resistance to the abrasive, rock-shod ice. But the pre-glacial gap had been developed on the line of a fault in the crust, so that deep gouging was possible – the out-flowing tongue of ice made up in depth what it lacked in width. Consequently, the pass – the Plas-y-Nant col in Fig. 80 – was converted to a deep trench, now occupied in part by the Cwellyn lake-basin and carrying the Gwyrfai across the line of the pre-glacial watershed.

A familiar locality is the gap between Nant Francon and the Ogwen valley (Plate 71). A view up Nant Francon shows the trough ending in a rock-step about 200 feet high, over which the stream from Llyn Ogwen descends in cascades. Downstream of the step, the sides of Nant Francon have been shorn away by ice. There is no difficulty in showing that a glacier moved down the Nant Francon trough, for scoured-and-plucked rock-knobs on the valley-floor are exceptionally well carved. But the rock-step is itself plucked on its downstream side. Ice in quantity evidently came through the pass at the head of the trough. Furthermore, rock-knobs on and near the shores of Llyn Ogwen indicate ice-movement towards the gap from the far side.

Within the Ogwen valley, drainage towards the Ogwen through the gap is separated from the easterly drainage of the Llugwy merely by an ill-marked hummocky band of glacial debris. For the first mile or two the Llugwy flows along the valley-floor with a gentle gradient, but towards Capel Curig it becomes incised into superficial deposits. The river-profile, projected upwards beyond

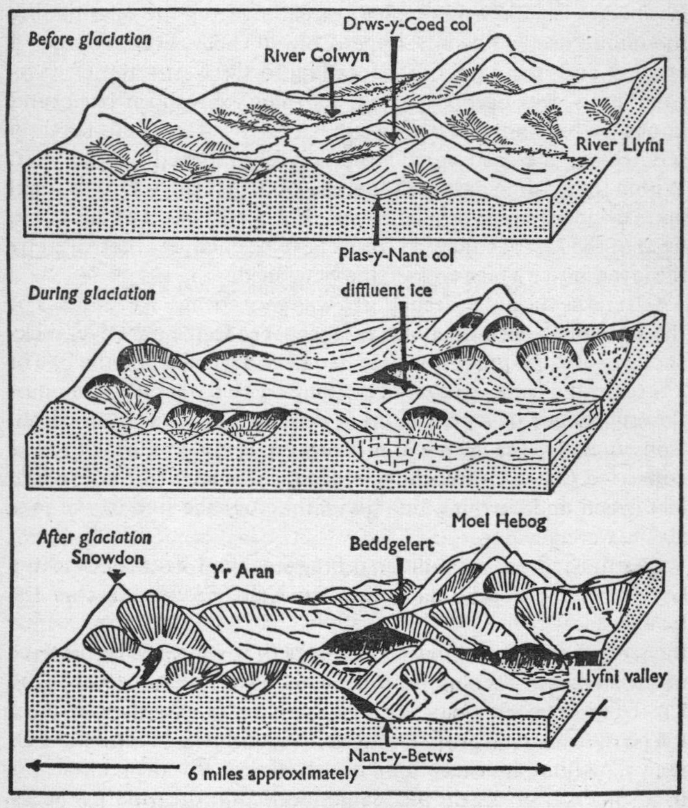


Fig. 80 Glacial breaching west of Snowdon

the low barrier which separates the Llugwy from the Ogwen, points directly into the big corries on the flanks of the Glyders (Fig. 81), which represent former heads of the Llugwy. There can be no

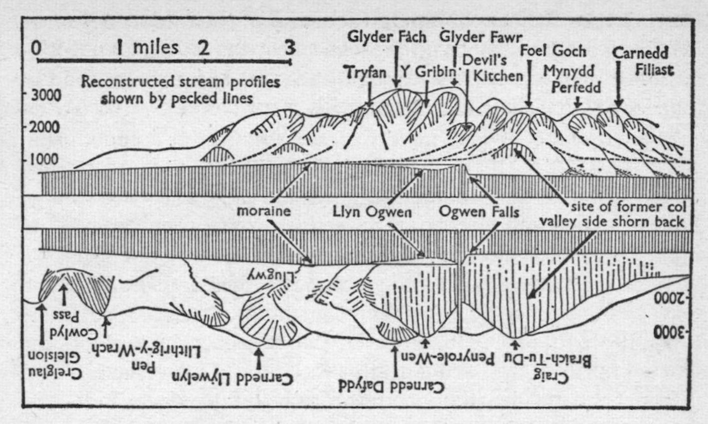


Fig. 81 Valley-side profiles, north Wales

doubt that the pre-glacial divide crossed the line of the Ogwen well to the west of the present line. It ran along the summits of the Carnedds, curving gently round to unite with the crestal watershed of the Glyders. The great breach was made by ice which, abundantly nourished from the mountains on either side, broke out in three directions. One route, running down-valley to the east, could not accommodate the whole discharge. Another went across the Carnedd ridge, where the Cowlyd pass was deeply scoured; but at the head of Nant Francon, on the side of a pre-glacial pass, a very deep breach was made and the watershed considerably displaced.

The Finger Lakes in west-central New York State result from the wholesale breaching, by sheet-ice, of a belt of sandstone hills. On their northern side, the hills overlook the borders of Lake Ontario, where the ground ranges in height from 300 to 500 feet, and is liberally bestrewn with drumlins. Ice, moving southwards from the Lake Ontario Basin and moulding the drumlins of the borderland, was thick enough to override the 2000-foot-high summits of the sandstone belt. Its erosive effects were greatest on the lines of pre-glacial valleys, which were deepened into the many basins now occupied by the Finger Lakes (Fig. 82). Lakes Cayuga and Seneca, the largest of the Finger Lakes, are each 40 miles long.

Respectively, they are 435 and 718 feet deep, their floors descending below sea-level. Beneath the lake-bottoms lie unknown thicknesses of drift material, for gouging was far more severe than is indicated by the depth of water alone.

Indeterminate divides now separate streams draining northwards into the heads of the Finger Lakes from streams flowing directly towards the Atlantic. Although the valleys now containing Lakes Cayuga and Seneca are thought to have been drained, in immediately pre-glacial times, by northward-flowing rivers, it is by no means certain where the pre-glacial watershed lay. The direction and extent of watershed-displacement cannot, therefore, be measured. It seems probable, however, that the lakes as a group are located north of, rather than athwart, the old watershed. Their present form and distribution appear to be influenced by morainic dams, such as have been proved to exist at the northern ends of some of the basins.

Glacial breaches made by ice-caps clearly illustrate the fact that ice-movement is only partially controlled by reliefs. Ice-sheets respond far less readily than do rivers to irregularities of the ground. A sheet of moving ice some hundreds – or thousands – of feet thick is not dependent on an integrated system of downhill slopes. It can cross watersheds, invade valleys, block the courses of rivers, and lay down deposits which obstruct post-glacial streams.

In lowland areas which were formerly covered by ice-caps, every

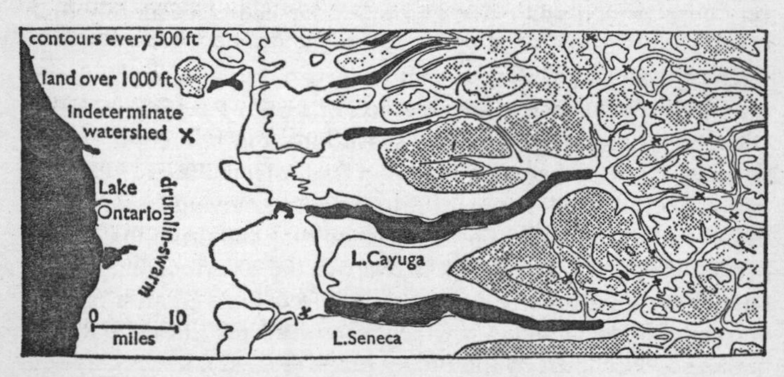


**Fig. 82 Some of the Finger Lakes**

gradation can be traced from minor and temporary derangements of the stream-system to the complete obliteration of the pre-glacial landscape. In East Anglia, thick deposits almost entirely conceal the pre-glacial topography. The pattern of post-glacial drainage can bear very little, if any, relation to the pre-glacial pattern. Similarly in the U.S.A., the drainage of what is now the Ohio valley has been drastically altered by glacial invasion and glacial deposition (Fig 83). General and severe erosion, as opposed to

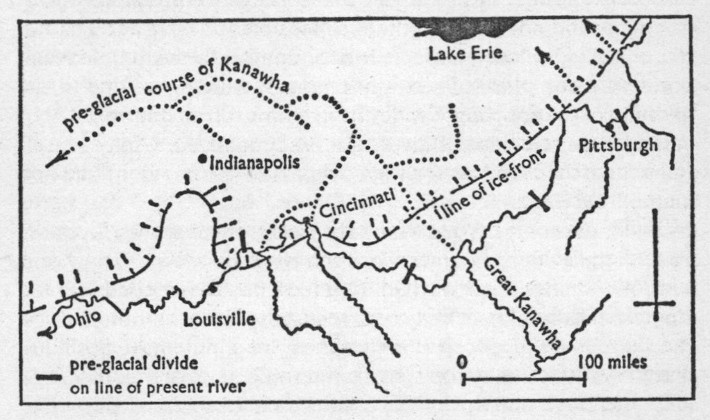


Fig. 83 Drainage-changes in the Ohio Basin (adapted from Flint and others)

general and abundant deposition, is also likely to give rise to a completely new set of streams, as in Finland and large areas of northern Canada (Plate 72).

Because their lines of pre-glacial drainage have been so completely obliterated, areas which have been severely scoured and areas which are thickly overspread with glacial deposits tend to be less interesting than areas where the details of diversion can be reconstructed. There are notable exceptions to this general remark, one of which will be described presently. But as a rule, there is far more scope for the investigation of drainage-changes in those areas which formerly lay at or near the ice-front than in places which were well within it or well beyond it at glacial maximum.

The complex history which has been worked out for the Thames system well illustrates this principle. The Kennet-Thames is the descendant of one of those master streams which, draining from west to east across Britain, can still be identified from their surviving remnants. The Kennet rises in an area which was subjected to intensely cold climate at glacial maxima, but which was not actually invaded by ice. The old valley of the lower Thames, east of Ware (Fig. 84), has been blanketed by ice-deposits and outwash, with great thicknesses of boulder clay representing even greater thicknesses of very dirty ice. In the absence of ice-invasion, glacial diversions did not occur in the Kennet valley. The lower Thames was entirely displaced from its former course. In the middle reach occurred a complicated series of changes, which were due to the invasion of the pre-glacial valley by ice coming from the north.

Before the sequence of diversions is summarized, it may be well to indicate the evidence for the flow of the Thames along the line marked by Henley, Marlow, St Albans, and Ware – that is, to establish the course from which the ancestral river was diverted. Part of the evidence comes from the reach between Henley and Marlow, where a northward loop of the Thames is incised into the Chiltern Chalk. The ice did not extend as far as this locality, and it is consequently inconceivable that the river should have been displaced *into* the Chalk country from a more southerly course. We infer, therefore, that the incised Chiltern reach represents part of an ancient course which the river followed when it flowed at a higher level – that is, at some distant period of its history. Now the Henley-Marlow reach is aligned from southwest to northeast. Projected northeastwards, it leads towards the Vale of St Albans. South of the Vale of St Albans lies a low plateau, capped by glacial deposits above river-laid gravels. Among the river-gravels are found pebbles derived from the Lower Greensand formation. The nearest exposures of Lower Greensand occur north of the Chilterns and south of the North Downs; one or other of these may be taken to have supplied the identifiable pebbles. There is no evidence for the previous existence of a river cutting southwards through the Chilterns, at least at the time when the gravels were laid down. It is therefore concluded that the pebbles came from the south – that is, from the lower Greensand of the Weald. But in that case there must have been a continuous gradient down which they could be transported – they could not have been carried into the Thames valley and up the other side. Thus it is concluded that, when the pebbles were introduced, the Thames in its present form did not

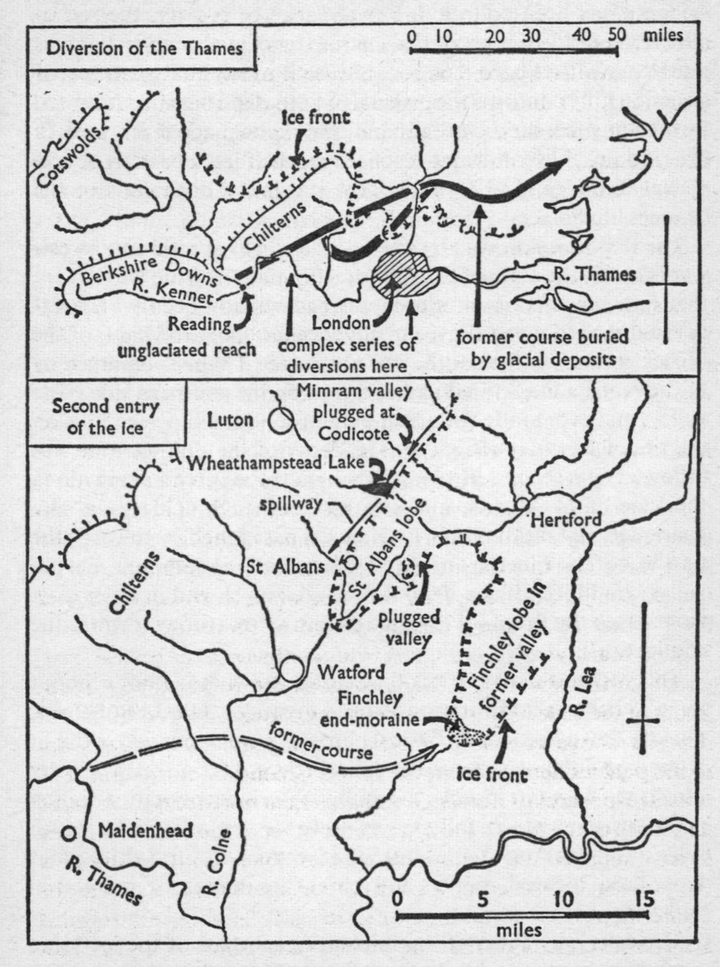


Fig. 84 A stage in the Diversion of the Thames (based on the work of Wooldridge and others)

exist. But the river coming from the west must have discharged by some route or other. That route lay along the Henley-Ware line.

Since this interpretation was first made, abundant supporting evidence has been collected. For the sake of brevity, the reconstruction of old courses of the Thames by means of river-terraces must pass with a bare mention. Suffice it to say that such reconstruction fully confirms the original hypothesis. The paths followed by invading ice, on the other hand, should be made clear, both in order to provide confirmation of the reconstructed pattern of preglacial drainage, and to show how successive diversions of the Thames took place.

The first invasion was by an ice-lobe which spread southwestwards along the Vale of St Albans. Although the Vale was not, at that time, as deep as it is today, it had already been widely excavated, and allowed the ice to advance far along the flank of the higher ground on the south. The obstructed Thames, dammed by the advancing ice, spilled through gaps on the southern side of its valley, and eventually found an outlet round the block of hills on the site of Finchley (Fig. 84). The decay of the first ice-lobe was followed by a second invasion, when one lobe spread again along the Vale of St Albans, and another reached Finchley – a sure guarantee that, at the time, a valley led past Finchley towards the northeast. The Finchley route was now finally abandoned; the old valley was blocked by a deposit of sand, gravel, and boulder clay, from which the Brent – a reversed stream – flows today towards the southwest.

This account does far less than justice to what is known about the diversions of tributaries of the Thames. A number of these, flowing down the slope of the Chilterns, were dammed by the St Albans ice-lobe, and forced to spill across cols in a southwesterly direction. At some points the spillways are deeply cut, forming great trenches which fall steeply towards the valley into which they once discharged the obstructed waters. Several drift plugs have been located, which still block the routes of pre-glacial valleys.

Deposits laid down at the fluctuating margins of the ice-lobes were hummocky and irregular. In the Vale of St Albans such deposits enclosed temporary lakes, in which varved (banded) clay was laid down. Not all is known about the number, size, extent, and history of these lakes, but the characteristic deposits are quite widespread, suggesting a time when surface drainage was far less organized than it has since become.

Lakes bordering the ice-front attained their greatest extents where ice advanced towards rising ground, so that large bodies of water could be impounded. In the English Midlands there was at one time an ice-dammed lake 750 square miles in area. Glacial invasion reversed the pre-glacial drainage and greatly displaced the watershed between streams flowing to the Bristol Channel and those flowing to the North Sea.

Before the third of the four complex glacial maxima of the Ice Age, the Warwickshire Avon did not exist. The area which is now the basin of the Avon drained towards the northeast, by way of a river Soar which was far longer than the present stream of that name. In other words, the piece of country in question was included in the basin of the Trent (Fig. 100). Ice, approaching from the northeast, north, northwest, and west, impounded surface water against the extremity of the north Cotswolds and their continuation in the Oxfordshire Hills (Fig. 85). The lake – Lake Harrison – spilled through gaps in the line of hills. It is certain that two gaps functioned as spillways, and highly likely that a third was also used. The first two are the gaps of Moreton and Fenny Compton; the third is the Watford Gap, which is now choked by outwash material.

For some time the level of the lake was controlled by the Fenny Compton Gap, a shore-platform being eroded by waves at a height of 410 feet above present sea-level. The platform can easily be distinguished in the field, and has been mapped in detail. It extends into the Fenny Compton Gap, disappearing in the valley of the Cherwell beyond.

The lake was eventually obliterated by the deposition of sediment and by the spasmodically advancing ice. When the ice started to decay, as a result of climatic change, the southwestern edge of the former lake-basin was first laid bare. A stream formed on the exposed lake-bed, draining towards the head of the Bristol Channel, and grew rapidly headwards to become the Warwickshire Avon as we know it today. Thus the drainage of the Avon Basin

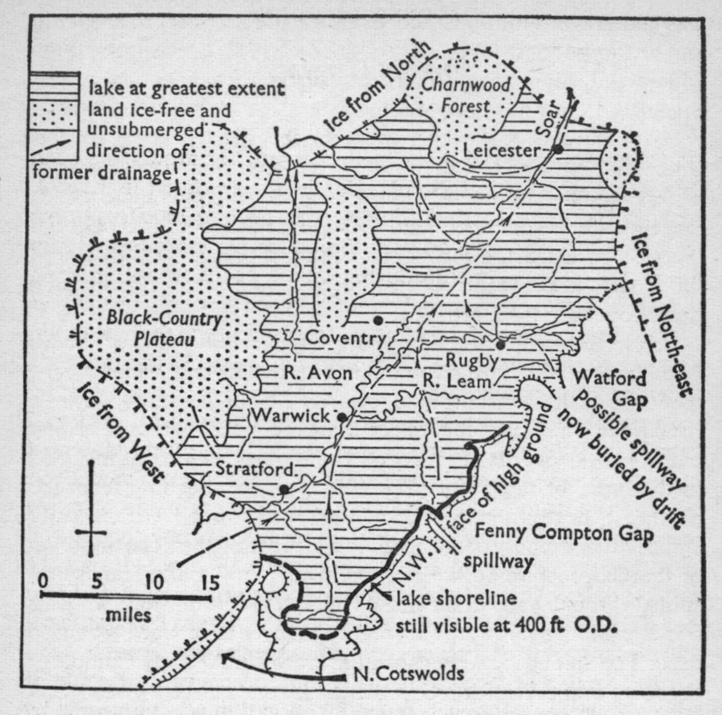


**Fig. 85 Lake Harrison (re-drawn after Shotton and Dury)**

has been twice reorientated through the agency of ice – first through gaps to the south, and then towards the southwest. The present river takes a line diametrically opposite to that of the pre-glacial stream which it has replaced.

CHAPTER FIFTEEN

# Frozen Ground

AT the Sveagruvan coal-mine, in Spitzbergen, the ground is frozen to a depth of 1000 feet. Similar depths are recorded in Alaska. At Nordvik, in the north of Siberia, frost reaches 2000 feet below the surface. We have to do with something far more severe than freezing of the waste-mantle during winter. In the lands bordering the Arctic Ocean, the ground remains frozen throughout the year – with the exception of the topmost few inches, which thaw out in summer. Permanent frost strikes deeply through the waste-mantle into the solid rocks beneath.

In combination with seasonal thaw at the surface, perennial frost in the subsurface exerts profound effects on the development of land-forms and on settlement and communications. These effects become especially noticeable when attempts are made to apply engineering techniques developed elsewhere. During the Second World War, very grave engineering problems arose during the driving of the Alcan Highway through western Canada and into Alaska. During the 1970s, the builders of the Alaska Pipeline faced still graver problems. Frozen ground has special properties.

Investigators of land-forms have long been aware of some of these properties. However, much investigation was pioneered by Russians – as was also true for the investigation of soils – and results published in Russian are not especially accessible to researchers in the Western world. Interest in the economic development of Alaska and northern Canada has enforced attention.

New words are needed to express new ideas. Among them, the linguistically deplorable term *permafrost* is coming into general use to mean (i) perennially frozen ground, (ii) permanent frost. By extension, it is also made to include the topmost layer which is subject to seasonal thaw. In the extended sense, permafrost can be divided into three classes – continuous, discontinuous, and sporadic (Fig. 70*b*).

In the belt of continuous permafrost, the subsoil never thaws. Temperatures at 30 to 50 feet below the surface are below 23° F, and go as low as 10° F in places. Continuous permafrost comes south of the Arctic Circle – about 1500 miles from the North Pole – in northern Canada, and farther south still, in Siberia, to points 2000 miles from the Pole. In the belt of discontinuous permafrost, the frozen ground is interrupted by unfrozen patches. Ground temperatures in the permafrost, between 30 and 50 feet down, range generally between 30° and 23° F. Sporadic permafrost, consisting of scattered patches of frozen ground in areas where most of the subsoil is unfrozen, occurs in a wide belt across Canada and crosses the Outer Mongolian border in the heart of Asia. Ground temperatures at 30–50 feet down are little below freezing in the sporadic permafrost.

Permafrost seems to be typical of near-glacial areas rather than of regions which are actually covered by ice. Although ice-caps prevent the ground beneath them from being warmed by the sun, they also prevent heat from being radiated into space. They act as insulators which obstruct the outward flow of heat coming from the earth's interior. This flow is admittedly small by comparison with incoming solar radiation, but is bound to militate against severe and deep freezing under thick ice. Significantly, regions formerly covered by ice-caps are not, on the whole, characterized by deep permafrost today.

Conditions favourable to permafrost include long, cold winters, during which the ground is chilled; short, cool summers, which are incapable of causing deep thaw; and low precipitation – especially low snowfall, for a thick snow-cover obstructs radiation from the surface of the ground. In other words, permafrost forms best where the climate is very cold but also very dry.

Data from Barrow, Alaska, exemplify some of the leading features of a permafrost climate. On the average, Barrow records 320 days a year with freezing temperatures. Mean July temperature is about 36° F, mean January temperature is below −20° F. Converted to its rainfall-equivalent, mean annual precipitation is little more than 4 inches – a very low figure indeed. About three-quarters of the total falls as snow, however, giving more than 30 inches of snowfall a year. But strong winds, averaging more than 10 miles an hour throughout the year, sweep away the loose snow and prevent it from forming a deep blanket. Frost strikes deep and hard.

Temperatures required for the formation of permafrost are uncertain, although an annual average somewhere between 24° and 30° F seems likely to be of the right order. At the present time permafrost is disappearing from some localities but spreading in others. On balance, it appears to have been in retreat up to about 1950, in accordance with a slight warming trend in the climates of middle and high latitudes. In Siberia, its southern margin retreated from about 1850 to 1950; but the movement may now have been reversed, matching a switch to a cooling trend.

At first glance, much of the terrain in the belt of continuous permafrost looks dull (Plate 73). Its relief is apt to be subdued, with wide gentle slopes merging smoothly into low hills. A thick mat of lichens, mosses, sedges, grasses, and low shrubs covers the formlessly undulating ground, failing entirely to break the monotony of the view. In detail, however, very sharp contrasts appear – particularly in the special forms assumed by rock-waste. But there again, just because one simple pattern can be almost endlessly repeated, the typical scene appears either monotonous or confusing.

Distinctive features fall into two classes – those due directly to frost-action, and those produced by the combined action of frost and of mass-movement. Frost tends to reduce rock-waste to small calibre, incidentally producing features which are described under the title of *frost-soils.* Mass-movement carries rock-waste downslope in enormous quantities.

Frost-shattering is not, of course, limited to the regions of permafrost, but it is likely to be unusually rapid there. Eminences based on solid bedrock are open to attack. On horizontal surfaces, and on gentle slopes developed on well-jointed rock, innumerable blocks can be detached. Strewn at random, and lying loose on the surface, they form block-fields (Plate 74). Single blocks can be wedged upwards, rising like a lift in a shaft.

The general product of frost-shattering is silt – i.e. rock-waste in which the limiting dimensions of single particles are $\frac{1}{16}$ and 1/256 mm. But in many parts of the permafrost belts, silt and angular rubble are intermingled. Year by year, the topsoil is stirred by frost. It is commonly found that the coarse material forms stone nets on horizontal surfaces and stone stripes on sloping ground (Fig. 86). Richmond has investigated nets and stripes in the Wind River Mountains, Wyoming, finding that nets are

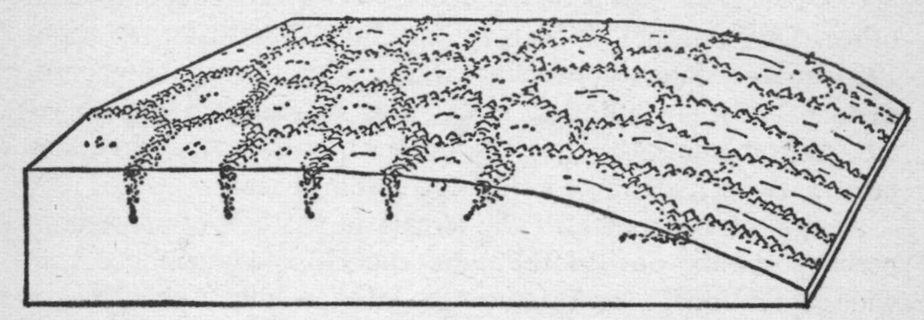

Fig. 86 Stone nets and stone stripes

confined to slopes of 4° or less. Each polygon has 4, 5, or 6 sides, measures 5 to 7 feet across, and encloses a mass of fine gravel, sand and silt in the middle. Below the surface the rock-waste is less well sorted. At 3 feet down, the polygonal distribution of fragments according to size has disappeared.

The dimensions cited can be taken as representative of the many available measurements of stone nets. Of the suggested explanations for polygonal ground, that favoured by Richmond seems the most likely. He points out that freezing waste has been known to crack suddenly and loudly, and that contraction-cracks often divide frozen mud into polygonal blocks. Although water expands slightly as it cools towards freezing point, ice contracts slightly as its temperature falls from freezing-point downwards. Hence Richmond infers that the formation of stone nets begins with the freezing, further cooling, and cracking of the waste-mantle. Ice-wedges form in the cracks. During summer thaw, when the material in the centre of each polygon expands, the surface is locally raised to a slight extent. The largest fragments fall into the hollows where the tops of the ice-wedges have melted.

Ground-ice, once established, can easily grow. Being crystalline, it is capable of enlarging itself in the manner common to mineral crystals in general. Temperature-changes below the level of freezing can also facilitate growth. For every 1° F by which temperature

53. Glacier trough containing lake, Switzerland

54. Glaciated highland, Jasper National Park, Canada

55. Corrie with moraine, Mt Biddle, Canada

56. Corrie glacier with bergschrund, Mt Andromeda, Canada

57. Glaciated rock-knob, North Wales

58. The Great Aletsch Glacier, with moraines and lake

59. Curve-over developed on formerly frozen ground, Northants

60. Boulder clay, Donegal, Ireland

61. Ice-scratched rock, Sweden

62. Stream falling from hanging valley, Alaska

63. The Sogne Fiord, Norway

64. Erratic of Silurian grit on Carboniferous Limestone

65. Structure of push-moraine, Herts

66. Varved (banded) clay from ice-dammed lake

67. Edge of Greenland ice-sheet

68. Eskers, Finland

69. Wave-eroded drumlins, Donegal

70. Glacial breach in watershed, Donegal

71. Glacial breach in watershed, North Wales

72. Ice-eroded lake-basins, Finland

73. Slopes developed on frost-soils, North Finland

74. Boulder-field, summit of Ben Nevis

75. Joints opened by frost, Cotswolds

76. Fissures opened in sandstone by frost, Northants

77. Cliff section at Pegwell Bay, Kent

78. Inselberg and pediment, Middle East

79. Rocky desert, Persia

80. Desert in Arizona

81. Systems of wadis in desert, Jordan

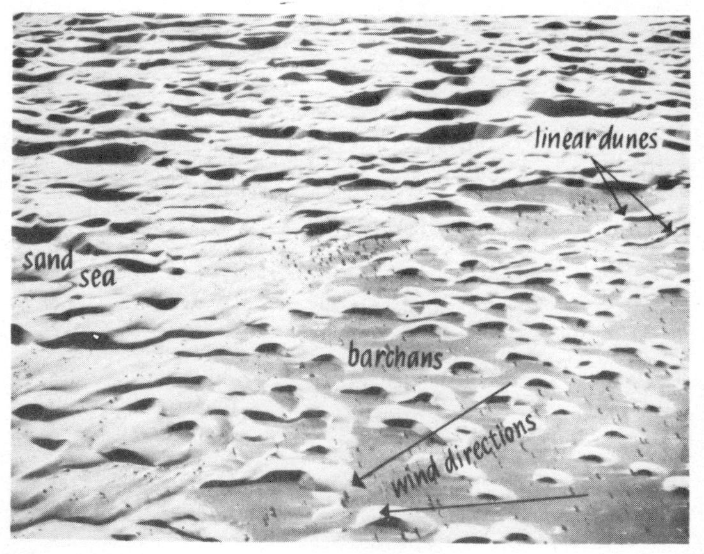


82. Dunes on the borders of California and Arizona

falls, the ice in a wedge of normal size undergoes a linear contraction of about 0·0001 inch. It seems probable that repeated contraction and re-growth can explain the formation of substantial wedges, and their partial survival during summer.

Ground-ice is known in forms other than wedges. Horizontal layers, lenses, and irregular masses are widespread, although it is understandably rare for them to be exposed at the surface. In some parts of the Arctic, blister-like hills dotting the generally even surface have been proved to contain thick lenses of ice. It is not yet certain whether the lenses have thickened in the normal course of growth, or whether their form marks a response to differential loading at the margins.

Seasonal thaw can go as deep as 10 feet in the belt of discontinuous permafrost, but 2 feet is the approximate limit in the belt of continuous permafrost, where no more than an inch or two may, in fact, thaw out each year. Thaw makes the surface layer very mobile – capable of moving on slopes of 4° or even of 1°. The waste is well lubricated by water, being always wet whenever it is not frozen. Summer temperatures never rise high enough to dry out the topsoil. At the onset of winter, the wettest material freezes the most readily, for its pore-spaces are filled with water instead of with air; water transmits heat 25 times as fast as does air, and ice 4 times faster still. Thus poorly drained rock-waste freezes quickly, both at the end of the summer and also during summer when night-temperatures fall. Every fissure which meltwater can penetrate is liable to be wedged open by ice – hence the tendency for rock-waste to be fine in calibre. Ground-ice, continuing to grow through the winter months, can release so much water during the following summer that the soil becomes supersaturated – its water-content exceeds the volume of pore-space.

Sludging downslope, rock-waste tends to mask the break between high and low ground. Distorted by sludging, its internal structures become festoon-like in section. Stone polygons, curving over the edge of horizontal surfaces, become elongated into stone stripes running downhill (Fig. 86). Widespread sludging is mainly responsible for the featureless aspect of many Arctic landscapes. In conjunction with frost-shattering wherever bare rock is exposed it seems capable of producing terrain which is generally subdued –

superficially similar to the end-product of the normal cycle of erosion, as demanded by the original scheme of that cycle (p. 68). Because sludging is so much more rapid than creep controlled by temperate climates, one may conclude that erosion in the permafrost belt is remarkably swift. But too little is yet known for the final products of permafrost processes to be defined with certainty. It has been suggested that long-continued frost-action could be responsible for features which, in middle latitudes, would be classed as erosion-platforms – for instance, the low-lying coastal bench known as the Norwegian strandflat.

Meanwhile, it is abundantly clear that the permafrost belt sets its own problems, not only of understanding but also of habitation. Buildings underlain by permafrost behave in peculiar ways. If the frozen ground thaws out, buildings subside on the yielding waste. Alternatively, foundation-posts are moved by frost-action. Among the devices used to counteract the various types of disturbance are large jacks, which enable buildings to be raised or lowered according to the seasonal shift of the foundations. But despite all attempts at correction, it is common to find the planking of wooden buildings sprung apart, with stout timbers protruding from the main structure. On continuous permafrost, the best technique is to keep the ground-temperature down. Elsewhere, it is easier to thaw out the subsoil and to keep it thawed.

As a final illustration of the distinctive nature of the permafrost belt, a brief description will now be given of the lake-basins on the Arctic Coastal Plain of Alaska. In the relevant area – more than 25,000 square miles in extent, and up to 100 miles wide – the extremely subdued landscape is dotted with thousands on thousands of lakes. All but the largest rivers freeze solid every winter, and permafrost lies a few inches beneath the surface of the ground. Large quantities of ground-ice are known to occur in the form of sheets and irregular masses, while ice-wedges underlie a polygonal network of shallow troughs. About three-quarters of the whole area is covered with lakes, ponds, or small enclosed marshes. The water-bodies range in length from a few feet to 10 miles. Their margins are irregular in detail, but their general plan is typically oval, the longest of them reaching widths of 3 miles or so. All which have been sounded are shallow. More remarkable than their number is their arrangement – nearly all of them have their long axes running about 15° west of north (Fig. 87).

Black and Barksdale, who have studied the area in question, conclude that the lakes are due to the thawing of permafrost. The shallowest existing lakes are directly enclosed by permafrost at the present day, and it does not follow that the permafrost is now being reduced. On the other hand, the waves raised by the winds which now prevail tend to change the alignment of the lakes from roughly north to roughly east. Minor irregularities in the shores are put down to thawing in summer, when the water rises to temperatures

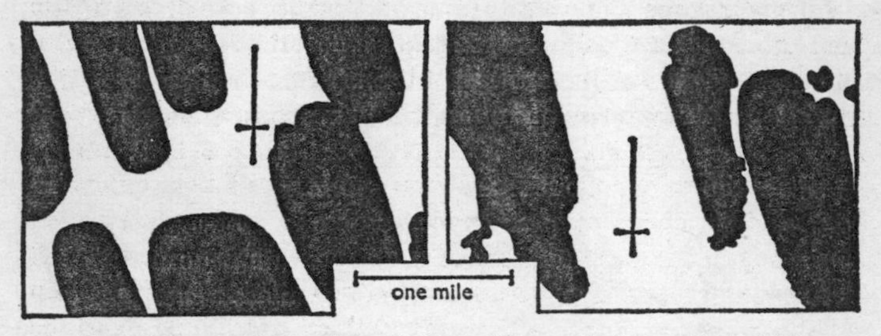


**Fig. 87 Some Alaskan lakes (adapted from Black and Barksdale)**

of 40° or 50° F and releases the immediate margins of the lakes from the grip of frost.

The current tendency towards a changed alignment implies a change in the direction of prevailing winds. A significant change in wind-direction implies a significant change in storm-tracks. The inference that such changes can be associated with changes in the distribution of permafrost agrees well with reconstructions of the climatic changes which accompanied maximum glaciation.

Certain extraordinary structures, first identified in the ironstone field of Northamptonshire, have been recognized as the effects of processes which operate in deeply frozen ground. Plate 59 exemplifies a type of scene familiar in the limestone hills of the south Midlands of England. The broad, gently sloping surface in the centre is developed on a resistant formation – the iron-bearing beds of the Middle Lias Marlstone. These beds, stripped of their former cover of weak Upper Lias Clay, underlie a smooth structural bench. Towards the southeast (towards the left in the picture) the height of the Marlstone declines, and the bench ends at the feet of modest hills which, composed mainly of upper Lias Clay, have protective cappings of Northampton Sand.

For all its remarkable smoothness, the bared surface of the Marlstone is not flat. It curves very gently over towards the north-west – on the right of the view – ending at the top of steep slopes which lead down to the head-valleys of the Warwickshire Avon. The curve-over is far too broad to represent the waxing slope described in Chap. 6, and in any case corresponds to a curvature of the underlying rocks. Considered in isolation, this gentle arching of the Marlstone is in no way unusual, for the rocks of this district are known to have been slightly flexured by crustal movement. But along the whole of its edge the Marlstone is similarly bent over towards the lower ground. In addition, the higher-standing North-

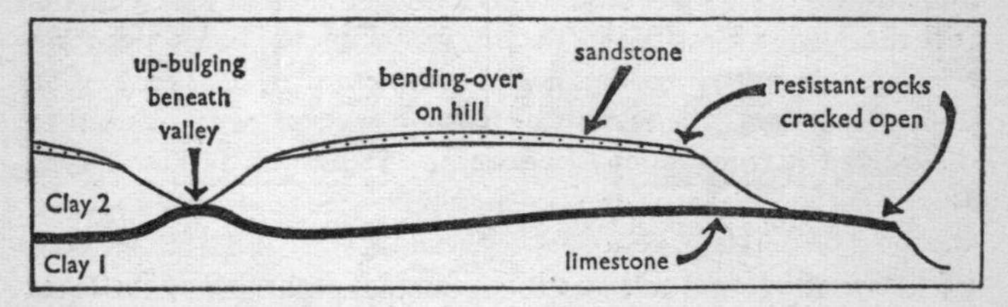


Fig. 88 Some structures produced in frozen ground

ampton Sand is also bent. The geologists who mapped the area about 100 years ago concluded that all the hills correspond to upfolds. They drew the logical inference that all the valleys correspond to downfolds. In actuality, however, the rocks beneath many valleys are upfolded, while the caps of strong rocks on the hills are bent down at their edges (Fig. 88).

Three clues point to the correct interpretation of this odd arrangement of structures. Firstly, wherever the Marlstone and the Northampton Sand are exposed in quarries on hill-brows, their joints are seen to gape wide open (Plate 75). The apertures are larger than might be expected as the result of unaided weathering. Secondly, large fissures are to be seen in some quarries; they are filled with the debris of overlying rocks (Plate 76). Thirdly, some exposures show that the whole of a resistant formation has been loosened. Its joints and bedding-planes have been sprung apart, and the spaces filled with clay; the whole resembles a brick wall, with the joint-blocks as bricks and the intervening clay the mortar. All these features are the result of freezing at depth.

Joint-planes and bedding-planes have been wedged open by ice. Because this ice was formed from dirty water, and because it melted irregularly, the fissures were partly filled with fine-grained material and were unable to close tightly when the ice finally melted. Ice-wedging doubtless played a part in the wide opening of joints on hill-brows, but an additional process was at work there. When the underlying clays thawed out, they became wet and weak. They were squeezed outwards by the weight of the cap-rocks, which sagged down at their edges and cracked apart.

Large masses broke away from the edges of the resistant cappings and moved downhill. A number of old slides have been located, but it is not always possible to distinguish a slipped block from one which has merely been detached by erosion from the edge of an arched-over cap. In places the moving masses were reduced to disordered rubble, as may be seen on the face of the Cotswolds. But broad cambering of the resistant rocks, rather than slipping, is the most typical feature to be observed. In the course of the bending, wide fissures opened on the curve of the arch: these, illustrated in Plate 76, are the *gulls* of quarrymen. Although they are primarily due to tension, it is quite possible that they were enlarged by the formation of vertical ice-wedges in the ground.

Cambering on the hills was accompanied by bulging in the bottoms of valleys. Like the clays of which the hills are mainly composed, the clays underlying the valleys were made mobile by freeze-and-thaw. Under the weight of the adjacent hills they were squeezed upwards. Bulging of valley-floors is to be compared with the upward creep of the floors of mine-galleries, and with the troublesome bulging which occurred when the Panama Canal was cut. Its cause is the differential unloading of weak rocks, which can be plastically deformed. Bulging is not confined to regions of frozen ground, as will be seen from the comparisons just made, but in the south Midlands it was chiefly – if not entirely – associated with thaw-and-freeze at depth.

Some rocks are far too strong to yield by bending. Nevertheless, they can be deeply shattered by frost-action. The Chalk on both sides of the English Channel is a case in point. At many places the topmost 50 feet or so seems to be well weathered, having been broken down into small joint-blocks. Its condition is due partly to normal weathering, but results mainly from frost-shattering in very cold conditions. One of the clearest illustrations may be seen at Pegwell Bay, in east Kent (Plate 77), where cliffs cut into the Chalk reveal sound rock at the base and shattered rock at the top. The angle of the cliff varies accordingly, the face being vertical in the sound Chalk and sloping in the zone of shattering.

At this same locality, old valleys exposed in the cliff-face are filled with a sandy Chalk rubble known as coomb rock. This is nothing but Chalky sludge which was mobilized by thaw-freeze. Above the sludge comes a layer of brickearth – a true loess, the nature of which has been proved by analyses of grain-size. Loess in Britain, as in Europe generally, is a sign of near-glacial climate. It was laid by powerful and bitter winds which, scouring the fans of glacial out-wash and the silt-choked valleys, raised thick storms of dust. The loess records cold, very dry climate. The underlying sludge was formed in a preceding episode of moister conditions when the topsoil thawed annually. Beneath the sludge comes the shattered rock, with the depth of shattering indicating the depth of penetration by frost.

At the last glacial maximum, a belt of loess-steppe struck westwards across the centre of Europe, just reaching southeastern England at its extremity. Between this belt and the ice-fronts lay a region of tundra (Fig. 89*a*), where wind-borne dust had little chance to lodge; any dust that may have fallen was incorporated in the frost-churned tundra soils. Furthermore, during the oncoming of glaciation the loess belt was itself subjected to thaw-freeze in damp conditions. Hence sludge deposits are commonly overlain by loess, as at Pegwell Bay.

It used to be thought that the European loess was deposited by easterly winds circulating clockwise round a high-pressure system above the Scandinavian ice-cap. But it is now known that the local distribution of the loess, its varying thickness, and the variations in its grain-size all indicate dominant westerly winds. Nevertheless, loess can form only in dry conditions; fossils of animals which lived in the loess belt are characteristically of the steppe type – antelopes, wild asses, and jerboas. Consequently the loess-sheets are ascribed to climates dominated for long periods by high atmospheric pressure over the European mainland. The frost-soils of the tundra belt seem to record the former extent of cold oceanic climate, which became progressively displaced from the interior as the ice-sheets built up.

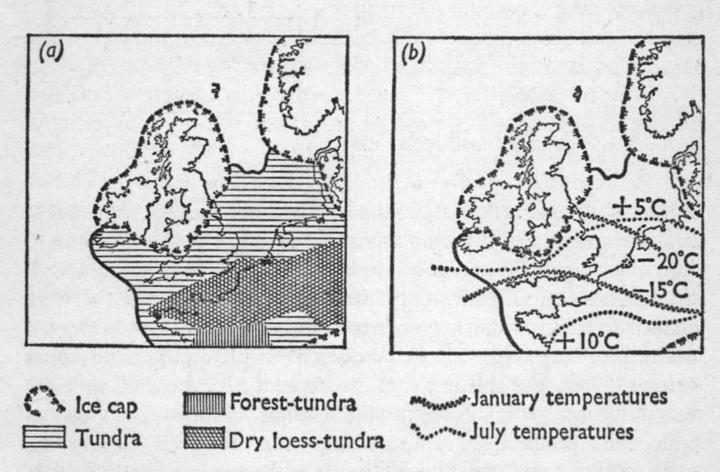


Fig. 89 Estimated vegetation (*a*) and mean temperatures (*b*) at the last glacial maximum (based on the work of Brusch, Büdel, Klein, Klute, and Poser)

Within the tundra belt, sludging was widespread and severe. To the coomb rock, formed on Chalk slopes, correspond other forms of sludge-deposit formed on rocks other than Chalk. The generic name for the whole group is *head*. Although the best exposures of head are to be found on cliffs, sludging took place inland as well as on the coast. Very large areas of the Paris Basin, northwest France, and southern England are known to have been affected. Old sludge-material, no longer subject to thaw-freeze, is known in the U.S.A. – e.g. in Montana. The shingle of the Patagonian desert has been re-interpreted as head. Probably originating with the sludging of fans of coarse alluvium on the lower flanks of the Andes, the

Patagonian head now lies perched on flat-topped divides between the present river-valleys.

In favourable circumstances, deposits of head can be dated. On Jersey, in the Channel Islands, great bulks of head are known on the north coast, where they rest on the 25-foot beach (Fig. 90*a*).

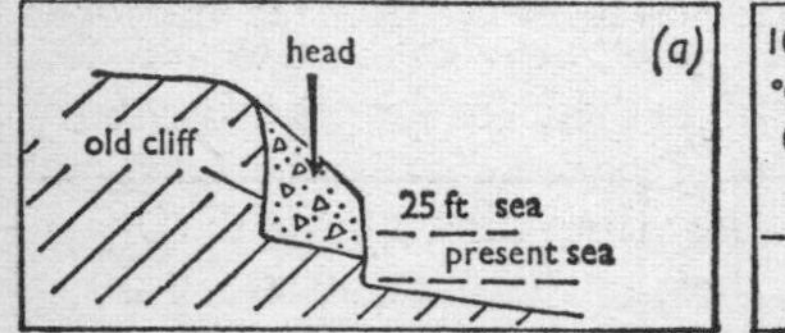


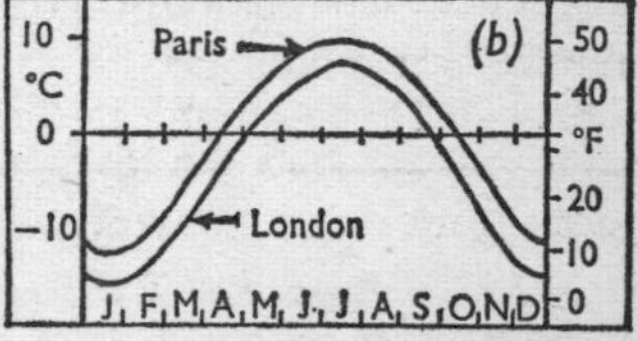


Fig. 90 (*a*) Cliff-section in Jersey
(*b*) Temperature-graphs for the last glacial maximum

The head obviously post-dates the beach. Since the beach was cut in the last interglacial, the head dates from the last glacial maximum.

While hillsides in general were likely to supply sludge-material, the greatest concentrations are found where the muddy flows were channelled along valleys. Numerous sludge-fans have been located where short but steep valleys debouch abruptly on to gentle slopes below. Where the shifting rock-waste had been layered in some way – where, for instance, it was originally a bedded deposit of water-laid sand and gravel – sludging often produced festoon-like contortions. At one time these structures were taken as evidence of submergence. They were thought to result from the rocking of stranded icebergs.

Loess-sheets and old frost-soils testify to cold conditions, but except in a general fashion they do not indicate the degree of cold. From what is known of existing areas of deeply frozen ground, a *minimum* estimate can be made of the severe climate which affected western Europe at the last glacial maximum. A more precise estimate, however, can be obtained by other means.

At glacial maximum, valley-glaciers were much longer than they are today. In any one large area, they seem to have been extended with rough uniformity. Similarly, the lower limit of permanent snow seems to have descended by a constant amount. Thus the average temperatures for areas which lay near the former ice-edge can be found by subtracting a constant figure from the present averages. The value used in subtraction is the difference in average temperatures between places on the former glacial limit and places on the present limit.

When present-day isotherms are adjusted in this way, a temperature-map for glacial maximum can be constructed (Fig. 89*b*). The July temperature indicated for London is well above freezing at about 7° C (45° F), but the January figure goes down to – 16° C (3° F). These values may be used as the basis for a graph (Fig. 90*b*), which suggests that the last glacial maximum gave the London district 7½ months a year with average temperatures below freezing. Corresponding figures for Paris are: July average, 10° C (50° F); January average, –12° C (10° F); 5½ months with average temperatures below freezing. These were the conditions in which the ground was permanently frozen at depth, and in which the topsoil thawed in summer.

In the 25,000 years since the last glacial maximum, the loess-steppe has vanished from west-central Europe and the tundra has migrated to the mountain-tops. A span of 25,000 years has been far too short to allow glaciated landscapes to be re-modelled by water. Old frost-soils, fans of sludge at the feet of hills, and sheets of head on numerous steep slopes, preserve some of the forms of near-glacial terrains. Hence the opinion that the land-forms of western Europe are abnormal (p. 71), in the sense that they are not due wholly to water-action. Those who consider the landscape of normal erosion to combine broad, gentle slopes with steep, abrupt hillsides naturally draw attention to the many signs of earth-sculpture in cold climates, for it is precisely the processes of freeze-and-thaw which are likely to produce subdued convex hills grading smoothly into the surrounding lower ground.

Research on fossil pollen has established the succession of climatic events since the last glacial maximum. The first recession of the tundra in favour of birch forest was partly offset by a return of cold, but the grass-sedge cover and the frozen ground beneath it were eventually displaced. Birch-pine, pine, and pine-hazel forests spread successively across much of northwest Europe, in climates which were colder, but also drier and more settled, than those of the present time. About 7000 or 8000 years ago came a rather sudden change. Western Europe was freely invaded by warm, moist air.

Mixed oak forest became extensive, with the moisture-loving alder and peat-mosses flourishing in the British Isles. In Sweden the hazel grew beyond its present limits, as was shown more than 50 years ago. At its maximum extent, hazel grew in places where the summers are now about 2½° C (4° F), too cold for it to survive.

Since that time further alterations have taken place. A decrease in rainfall arrested the growth of peat-bogs and restored some of the forests, but about 2500 years ago a rapid deterioration set in, when a new increase of rain was accompanied by increasing cold. The present climate is perceptibly cooler than that of 7500 years back. For this reason, the period of maximum warmth is called the Climatic Optimum. Usually, indeed, it is called the Post-glacial Climatic Optimum, implying that the Ice Age ended with the decay of the last great ice-sheets.

There is, however, no reason to assume that the Ice Age has ended. Ice-sheets are still with us, even though they have gone from the mainlands of Europe, Asia, and North America. It is quite possible that the so-called Post-glacial Climatic Optimum was the peak of an interglacial. Fossil soils from the last interglacial period strongly resemble those of today, both in type and in distribution. It follows that the present climates are, broadly speaking, of an interglacial kind. The record of climatic changes in the last interglacial period, as reconstructed from the indirect evidence of vegetable remains, strongly resembles the succession of changes in the so-called post-glacial period. It continues, however, through worsening conditions towards the succeeding (last) glacial maximum.

If we have reached, and passed, the peak of an integlacial, the present short-term tendency for winters to become milder is but a slight disturbance of a long-term trend, and another glacial maximum is due in about 10,000 years' time.

CHAPTER SIXTEEN

# *Wind, Sand, and Torrents*

VAGUE mental images called up by the word *desert* are apt to resolve themselves into pictures of sand-dunes, with a group of wandering Arabs and a few camels in the foreground, and a patch of date-palms in the distance. In actuality, desert scenery is extremely variable. Only a quarter of the world's extent of desert country is sandy. Bare rock and spreads of gravel account for the other three-quarters (Plates 78, 79). Gravel deserts look much the same wherever they occur, but rocky and sandy deserts display very marked internal contrasts in addition to being unlike one another. Desert vegetation also ranges widely. Some desert areas support no plants at all. At the other end of the scale come parts of the North American desert, where an admittedly specialized but nevertheless impressive set of plants can grow (Plate 80).

All the same, deserts have enough in common to justify discussion in general terms. A practical point is that little is known about most of them. Consequently, it is difficult to support mere description by explanation and so to supply the reasons for scenic differences. More important, all deserts are dry. They receive too little rain for plants to grow in a continuous mantle. Even at its thickest, desert vegetation is patchy or scattered. It follows that desert soils are incompletely developed. Loose desert sand can be called soil only by courtesy. Unaffected by plants or by soil bacteria, it falls into the limiting class of mineral soil. Even where desert plants grow, and where rock-waste can be acted upon by organic processes, the soils remain in a state of arrested development. Permanently short of water, they can never be transformed into soil as the term is understood in rainy areas.

In the heart of desert regions, the distinctive qualities of desert climate are unmistakable. Drought, sun, wind, occasional rainstorms, and heat by day characterize the climatic year. Things are different on the desert margins.

Unless a desert region is bounded by highland, desert climate and desert scenery vanish by replacement. In such circumstances, the status of a given locality may be very doubtful, and the limits of the true desert very difficult to fix. In some areas the matter is complicated by the effects of past climatic changes – the Mediterranean borders of the Sahara were perceptibly rainier, two or three thousand years ago, than they are today. Desert climate has invaded regions which were formerly not more than semi-arid, and semi-arid climate has encroached on regions which were truly humid. Too little time has elapsed for the scenery to be fully adjusted to the changes, so that the scenic and climatic limits of the existing desert do not coincide.

Attempts at defining dry climates rely ultimately on the relation between evaporation and rainfall. It can be urged that dry climates begin where evaporation is powerful enough to remove all the water which falls as rain. Inside a boundary so drawn, evaporation is, by definition, excessive. It could dispose of more rain than actually falls. In the context of earth-sculpture, the idea of a potential excess of evaporation is highly convenient. It implies that the relevant climates are too dry to nourish permanent rivers.

This is not the same as saying that no permanent river crosses any desert. It is possible, although rare, for rivers supplied by copious rain or snow to invade deserts, cross them, and reach the sea. Two very well-known examples of such rivers are the Nile and the Colorado. But streams originating within dry areas fail to break across the regional limits. Large invading rivers can also be destroyed by evaporation. The Jordan ends in the salty waters of the Dead Sea, the Syr Daria and the Amu Daria feed the inland Sea of Aral, and the Tarim enters the marsh-rimmed lake of Lop Nor. In really dry areas, interior drainage is the rule.

If the principle is accepted that dry climates begin where evaporation is equal to rainfall, it ought to be possible to establish some kind of numerical limit. Allowance has to be made for the fact that 5 inches of rain falling in the cooler season (if any) are more effective in wetting the ground than are 5 inches falling in the hottest part of the year. Rainfall totals alone are not enough to show where desert climates begin.

In practice, moreover, regions of semi-arid climate are distinguished from regions of truly arid climate. The title semi-arid implies a deficiency of rainfall for the year as a whole, with serious and prolonged drought part of the time, but leaves room for a season when considerable rain falls. Hence the landscapes of semi-arid regions include prominent erosional features which have been modelled by running water. Even in parts of the deserts proper, water leaves its mark.

In one well-known scheme of climates, the outer limit of dryness is set where the annual total of rainfall, in inches, is equal to 0·44 ($t$–19·5), $t$ being the mean annual temperature in ° F. If at least 70 per cent of the rain comes in the hot season, the formula becomes 0·44 ($t$–7), while if at least 70 per cent comes in the cool season, the expression 0·44 ($t$–32) is used. In this way, allowance is made for the low effectiveness of hot-season rain and the high effectiveness of cool-season rain. The limit so determined is the outer limit of semi-arid climate. True desert climate is considered to begin where the rainfall-total is down to half the figure calculated for the outer boundary of semi-arid climates. Rainfall-totals corresponding to selected average temperatures are shown in the following table:

| Mean annual temperature, °F | 80 | 70 | 60 | |
|---|---|---|---|---|
| Approximate mean annual rainfall (inches) at the outer margin of semi-arid climate | 21 | 16½ | 12¼ | winter concentration |
| | 26½ | 22 | 17¾ | even distribution |
| | 32 | 27½ | 23¼ | summer concentration |
| Approximate mean annual rainfall (inches) at the outer margin of arid climate | 10½ | 8¼ | 6 | winter concentration |
| | 13¼ | 11 | 9 | even distribution |
| | 16 | 13¾ | 11¾ | summer concentration |

Baghdad, with a mean annual temperature of 73° F, and 7 inches of rain a year concentrated in the cool season, falls squarely within the climatic boundary of desert; for 0·44 (73–32) is about 18 inches, and half that value is 9 inches.

Refined calculations can be made when the influence of transpiration is reckoned in. For many purposes it is highly desirable to allow for water-losses through the substance of plants in addition to, and as opposed to, losses by evaporation direct from the ground. It can be shown that the maximum combined loss by evaporation and transpiration to be expected at Sante Fé. New Mexico, is equivalent to some 24½ inches of rainfall. The actual fall is 14½ inches. Taking into account the distribution of rainfall through the year, these values indicate a semi-arid climate for Santa Fé. Fresno, California, with less than 5 inches of rain, could do with more than 25 inches. Its climate ranks as arid.

Whether a particular climate is semi-arid, with serious deficiencies of water during part of the year, or whether it is truly arid, with water chronically short, the amount of rain which actually falls can be surprisingly high. Semi-arid regions can record totals which, in cooler parts of the earth, would mean quite wet climates. Thus although most desert streams are intermittent, surface-water is capable in some localities of dominating the processes of earth-sculpture. In semi-arid regions it can be generally dominant. These statements apply with all the greater force because rainfall in dry climates is irregular. When it does come it can be violent.

It has already been mentioned that large portions of the African plateaus are typified by pediment-and-inselberg landscapes. The pediments appear as broad, gently sloping surfaces cut across solid rock. They lead from the feet of mountains or isolated hills down to the rivers, decreasing in gradient downwards. Inselbergs are sharply upstanding hills, the remnants of once-larger masses. Steep mountainsides and the flanks of some large inselbergs are slit by gullies, showing that channelled streams share in the work of erosion. But weathering of the whole surface, and the downhill movement of weathered rock, seem principally responsible for the wearing back of steep hillsides. By contrast, the pediments are open to the unchannelled wash which some regard as effective in shaping them. Their mode of origin, however, is not yet finally explained. Many valley-bottoms and larger depressions in semi-arid regions are heavily encumbered with alluvial deposits, carried downslope and lodged on lower ground. But the rivers seem capable in the long run of moving all the rock-waste which the floods bring down. Here is a fundamental contrast between semi-arid and desert regions. In deserts, where external drainage is most uncommon, land-sediments accumulate progressively. Only if their grade is very fine, so that dust can be blown away in very great amounts, can sedimentation be kept under control.

In deserts proper, the general lack of external drainage implies that sediments accumulate progressively. A useful distinction can be made between water-built and wind-built features. A parallel distinction is needed between features modelled by water-erosion, which resemble the water-eroded forms of semi-arid areas, and features shaped by the natural sand-blast of true deserts. When the contrasts among the classes of rocky, sandy, and gravelly deserts are added, it becomes possible to reduce descriptions of desert landscapes to order.

It may well be that variations in the abundance and grade of sand explain the contrasted developments of desert surfaces. Winnowing concentrates dune sand within remarkably narrow limits, for some 80 per cent falls within the range of 0·5 to 0·125 mm. diameter. These values place it chiefly in the classes of medium sand and fine sand. But Bagnold maintains that, although desert weathering is highly effective in breaking exposed rock, temperature-changes are unlikely to split particles smaller than 10 mm. across. He also finds that grains considerably bigger than normal desert sand merely bounce off solid surfaces on impact. It seems likely, therefore, that neither desert weathering nor wind-driving can be fully responsible for the observed grade of desert sand, even though individual grains are known to become rounded and frosted in desert conditions.

One must infer that most of the sand is provided ready-sized, as it were, by the breakdown of rocks of appropriate texture. Since desert sand is composed dominantly of quartz grains, it seems likely that sandstones contribute the bulk of it. Composed of grains crushed small on the beds of rivers or on the seashore, sandstones are capable of providing the fine-grained and medium-grained quartz sand which characterizes sandy deserts. Fine-textured crystalline rocks and desert torrents presumably make a supplementary contribution.

A particular desert cannot become sandy if loose sand of suitable grade and sufficient bulk is not provided. Very coarse-textured rocks weather into fragments too large to be freely moved. Gravel deserts – the gibbers of Australia – develop on weathered conglomerate, forming a kind of armoured surface which is highly resistant to abrasion. Individual fragments are usually coated with an enamel-like covering of oxides of manganese and iron.

If very fine-grained rocks disintegrate, their waste can be swept away entirely. In any event, rapid stripping either by the wind or by surface-water reveals the structures of the solid rocks in full starkness (Plate 79). Unsoftened by a cover of waste, soil, or plants, the land-forms of rocky deserts are so clear-cut that by comparison with the landscapes of rainy regions, they seem exaggerated.

Surface-streams, coursing intermittently down steep desert slopes, cut gullies similar to those noted for semi-arid regions. If rainfall suffices, whole integrated systems of valleys can be produced (Plate 81), even though no streams flow in them for months or years at a stretch. When rain does fall, runoff is extremely rapid. Unchecked by vegetation or by retentive soil, water pours into the valley-bottoms. Streams rise in torrential flood. Discoloured by mud and sand, they surge down-valley as long as the water lasts. But sooner or later they either die of evaporation or vanish into the loose thickness of rock-waste beneath. The largest and most frequent streams, transporting boulders and cobbles in addition to sand, build massive fans where their valleys debouch on to low ground. Enough water reaches the centre of some desert basins to create lakes, some long-lived and moderately salty, others very salty and quick to evaporate.

Pediments are known from the margins of many desert basins. Like the pediments of semi-arid regions, they are difficult to explain completely: but they do exist, and in great number. Many of them are traversed by rills (Fig. 91), but some rill systems may be relict. Desert pediments, unlike numerous pediments of semi-arid country, are liable to be partly or even wholly buried. Fans of alluvium spread across them from the mountain-foot, while the deepening sediments in the centre of the basins encroach steadily on their inner borders.

Desert sand, constantly stirred by uneasy winds, is highly abrasive. A wind reaching a speed of 11 m.p.h. at a height of 4 inches above the ground sets ordinary desert sand in motion. A wind of 22 miles an hour exerts a force equivalent to more than 40 times their weight upon grains which bounce up from solid obstacles. Consisting chiefly of quartz, wind-driven sand is hard. But although dust particles can be carried to great heights during storms, little of the sand rises more than a few feet above the ground. Sand-abrasion is concentrated at low levels, seldom taking effect at heights greater than 18 inches.

Pedestals, reminiscent of sea-stacks sapped at the base by waves, are the most striking results of undercutting. Exposed rock-surfaces undergo selective abrasion, which again can produce minor erosional features very like those visible on the seashore.

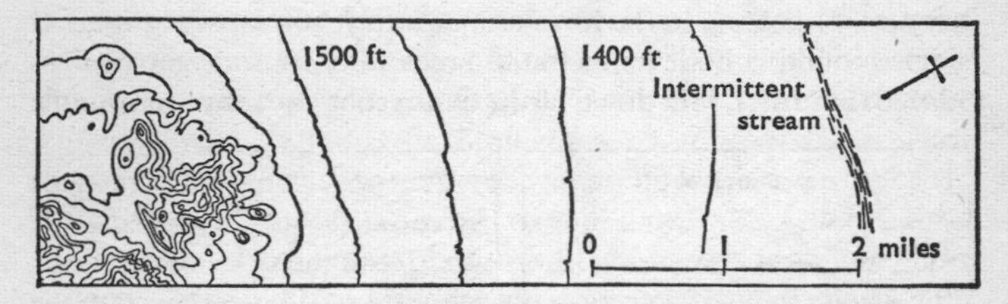


**Fig. 91 Inselberg and pediment, Sacaton Mts district, Arizona**

Hollows in rock-faces, ranging from tiny pittings and honey-combing to large but shallow caves, can however occur well above the range of sand-blasting. Often wrongly ascribed to this process, they are the products of cavernous weathering, a set of processes not yet fully understood.

*

According to Bagnold, sand-built features in the desert can be divided into five main classes. Sand-shadows and sand-drifts make up one class. True dunes, subdivisible into two groups, constitute another. The remaining three classes include whalebacks, low large-scale undulations, and sand-sheets. Each of the five classes marks a distinctive kind of response to variations in wind-speed, wind-frequency, wind-direction, abundance of sand, and local surface conditions.

It is commonly thought that dunes form where the flow of sand is obstructed by some fixed object such as a building or an upstanding rock; bushes and dead camels are also favoured. But this idea is as completely mistaken as the notion that obstacles cause meanders in stream-channels (p. 95). In actuality, desert dunes come most easily into being upon flat erosional surfaces which are devoid of large obstructions, and where the climate is so dry that rainfall and vegetation exert little or no influence upon the flow of sand. Sand deposited on the lee side of obstructions forms drifts and shadows, which, unlike true dunes, are destroyed if they move away downwind.

Where the wind is deflected round and over an obstruction, sand falls out of the driving air on the lee side. At the outset it forms two tongues, projecting in the form of a swallow-tail, but eventually a single mound is built up as the tongues thicken and merge (Fig. 92*a*). Sand-drifts, laid down at the exits from gaps through which

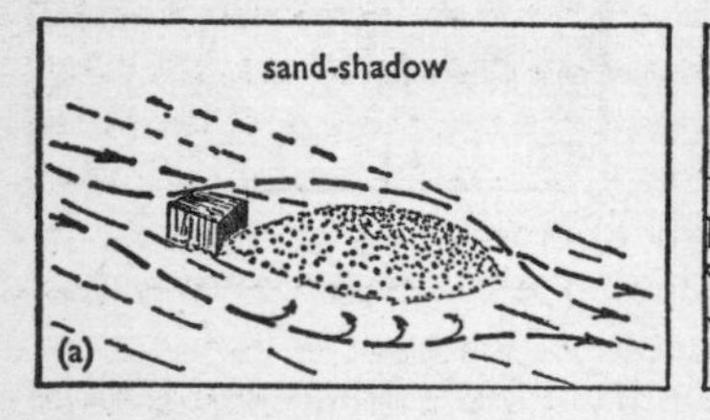


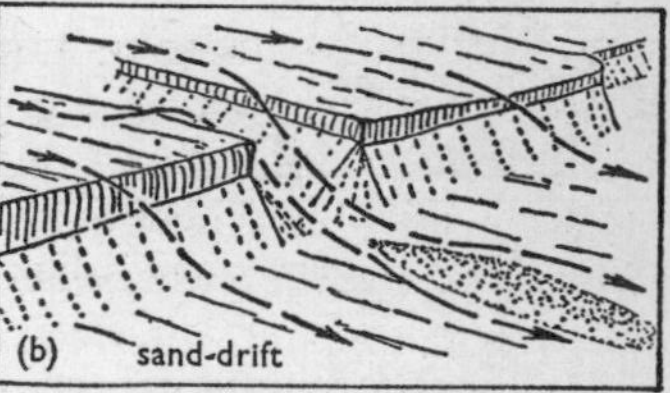


**Fig. 92 Sand-deposits near obstacles (freely adapted from Bagnold)**

the wind is funnelled, tend to be greatly elongated downwind (Fig. 92*b*). Neither shadows nor drifts resemble free dunes, either in plan or in profile.

The big swelling mounds called whalebacks consist principally of grains coarser than the average for dune sand. Bagnold interprets them as the relics of linear dunes. As will presently be seen, coarse sand is locally concentrated on the lower flanks of dunes, and the greatest concentrations are to be expected where linear dunes attain great lengths. At the same time, it may well be that whalebacks are to some extent comparable with the sandbanks of a braided river-channel, for these also begin with the accumulation of coarse material.

Whereas the whalebacks consist of bare sand, unfixed by vegetation, large undulating mounds composed of sand of average calibre can rise where patchy vegetation grows. Grass-bearing surfaces act as sand-traps. By reducing wind-speed very close to the ground, they increase drag. Falling sand forms a loose surface above which drag is still further increased. The plants, struggling upwards through the thickening deposit, maintain a retentive surface and allow the mounds to grow higher than their immediate surroundings. Mounds of this kind may be comparable to sandbanks in a braided channel which have been colonized by plants. They are necessarily confined to desert margins, where enough rain falls to support grass.

Little is known about the origin of wide, flat sand-sheets. However, it is at least possible that some of them are composed mainly of sand below a critical grade of size – sand fine enough to resist wind-erosion unless it is disturbed by something other than the wind. Grains of the relevant calibre are carried in suspension, being unlikely to fall to the ground unless the wind drops. Thus, if they do fall, they descend in an even powdery sheet, forming a surface which behaves as if it were smooth. Once laid down, they resist erosion when the wind rises. Bagnold finds that particles smaller than about 0·03 mm. diameter cannot be swept up again individually unless the surface is roughened in some way.

Most true dunes of the desert are recognizable either as barchans or as seif dunes (Plate 81). Barchans are crescentic in plan, with steep slopes on the lee side. They do not exceed some 1250 feet in width or 100 feet in height. Seif dunes are long ridges of sand, up to 300 feet or more in height and with lengths ranging up to 60 miles or even above. Seif dunes occur in huge families, each ridge parallel to the next. The distance between crests varies from area to area, being as little as 60 feet in some localities but as much as 1500 feet in others. In any one group of seif dunes, however, it varies little. With the instructive exception to be noted later, barchans and seif dunes are not found in association.

Barchans cannot be constructed unless a heap of sand rises to at least a foot. Only in these conditions can there develop the steep lee face which is a fundamental characteristic of the barchan. With smaller heights the grains can advance indiscriminately, but with greater heights a dune can increase in height, assume and retain its typical form, and advance across country.

These statements are based on the observation that sand can be carried by the wind in three ways. Fine grains carried in suspension have little bearing on the present topic. Coarse grains creeping across the surface under strong winds, and normal dune sand creeping under modest winds, have no more than limited relevance. The characteristic motion of blown sand is a kind of jumping, called *saltation* (Fig. 93a). The trajectories of individual grains always vary with the size and weight of the grains and with the speed of the wind, while the angle of take-off is influenced by the nature of the surface – grains bounce off pebbles, sometimes to

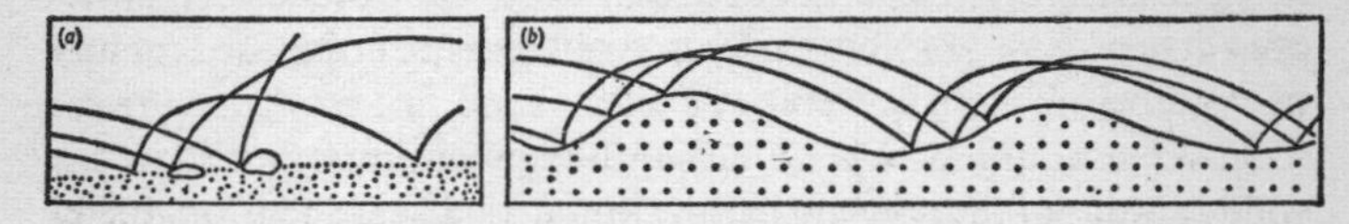


Fig. 93 (*a*) Leaping of sand-grains
(*b*) Grain-flights and ripples (both re-drawn after Bagnold)

heights well in excess of the 4 inches which is the average figure or of the 3 feet above which few sand-grains rise. But despite these complications, the idealized trajectory drawn in Fig. 93*a* is a fair representation of the average response of desert sand to wind-driving.

Now if a heap of sand is so small that grains blown off it fall beyond its limit, it cannot grow larger. In these circumstances ripples form but barchans do not (Fig. 93*b*). If, however, the heap grows above the critical height of one foot, some of the leaping sand is carried over the leeward face. The lower part of this face is protected from deposition. The upper part, however, continues to receive grains which are near the end of their trajectories. Thus the angle of slope is increased. When the angle reaches 34° the sand slips down across the whole face, so that the foot of the dune advances. Further deposition at the top, and further slipping, enable the advance to continue as long as the supply of sand holds out.

The main problem, then, is to discover how loose sand can be heaped high enough to promote the first development of a barchan. As Bagnold says, any change in the rate of sand-flow provokes a fluctuation downwind. A rising wind crossing a patch of sand does not immediately become fully loaded with sand-grains, nor is the general balance between the take-off and landing of leaping grains attained by a steady increase in the amount of sand in motion. Experiments with wind-tunnels show that, at the point where the flow of sand first reaches the rate at which it will eventually be stabilized, the air-speed close to the ground is above the speed to which it will eventually be reduced by drag. Consequently, sand is removed over a distance corresponding to the time taken by the wind to respond to changing conditions. Downwind, where the sand-flow is increasing, the drag is also increasing. The wind-speed is reduced, and more sand is deposited than is removed. In this way the loose sand-surface can be made uneven.

Rippling of the whole surface appears to be independent of the processes described above, except that it too is due to wind-transport. Bagnold finds that the wavelength of wind-ripples is closely similar to the length of trajectory of individual grains, and concludes that the leaping-distance controls the spacing of ripples (Fig. 93*b*). But dune-building requires heaps of sand larger than ripples. Unless sand can be locally concentrated, dunes cannot form. They seem to grow on patches of sand not less than 15 feet or so across, for smaller patches are stripped away before increasing load can be reflected in heavy deposition to leeward.

Dunes never form on flat expanses of fine sand, unprotected by larger grains. But where the sand varies in grade, or where a rocky surface is strewn with pebbles and large sand-grains, local concentrations are possible. Sand is trapped when gentle winds blow, and when the wind rises is collected by already-sandy patches which exert a powerful drag on the passing air. Bagnold stresses the important distinction between the winds which bring the sand and those which actually raise the dunes. In Fig. 94*a*, sand is transported by prevailing northerly winds which are usually gentle. Strong southwesterly winds concentrate the scattered grains on particularly sandy patches, where the height of 1 foot is exceeded and where barchans develop.

A variant process is that whereby sand is concentrated not in patches but in strips. If the surface of the ground is already striped by alternating bands of sand and pebbles, gentle winds are retarded over the pebbly strips where they deposit part of their load. With strong winds, on the other hand, drag is greatest on the sandy strips. Grains are removed from the pebbly bands and deposited on the strips which are sandy to begin with. Even if the surface is uniform, it encourages strip-formation if it is uniformly rough. Strong winds blowing across it become transversely unstable. Once above the critical height, a strip can continue its upward growth.

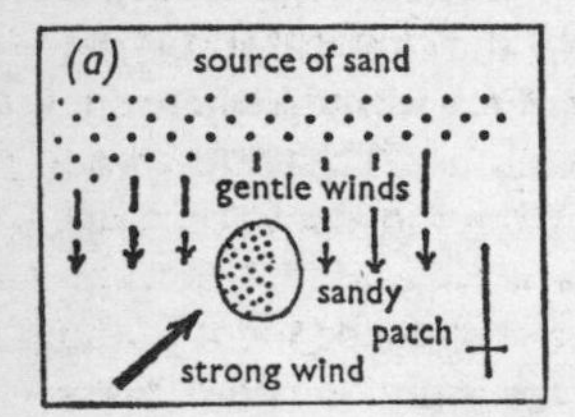


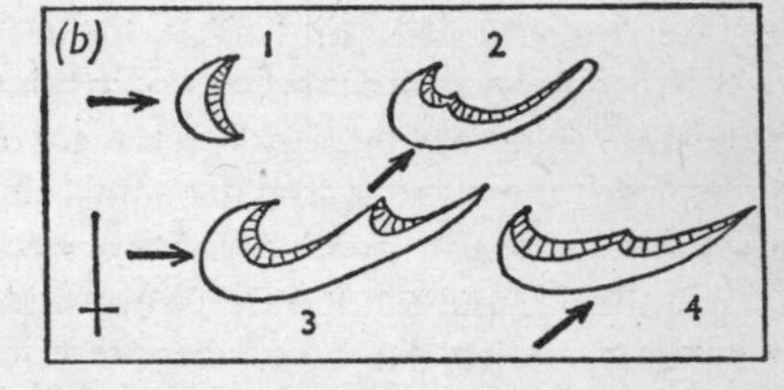


Fig. 94 (*a*) Mound-building
(*b*) Seif-building (both re-drawn after Bagnold)

Bulges on the downwind side bud off as independent barchans.

Once in being, a barchan migrates downwind, rather as an off-shore bar moves towards the land. Just as the seaward face of a bar is liable to be combed down, so the sand on the windward side of a barchan is winnowed away. But just as the landward face of a bar is renewed by material washed over, so the leeward face of a barchan is supplied with sand blown over the crest. The horns of the crescent project downwind, so that the leeward face is curved in plan, but the whole form represents an equilibrium-state.

Small barchans travel faster than large ones. A barchan which has escaped, as it were, from the locality where it was first built can last only so long as the wind which drives it forward carries enough sand to nourish it. Generally speaking, escaped barchans seem to be racing against extinction. At the other end of the scale, the largest barchans move very slowly indeed. Bagnold quotes Bead-nell's observations on barchans ranging from 12 to 60 feet in height to show that speed of advance decreases rapidly with increase in size. When the ratio of advance to height is drawn as a graph, and the graph extended, it seems likely that barchans exceeding some 115 feet in height ought to be stationary (Fig. 95*a*). The value reached in this manner agrees closely with the maximum heights recorded for actual barchans.

Seif dunes differ from barchans in being essentially linear. They are also elongated roughly in the direction of the wind, whereas barchans are transverse to the wind. But with seif dunes, as with barchans, it is important to recognize the distinction which may be necessary between the gentle winds which bring sand and the strong winds which build dunes. Discussion is handicapped by

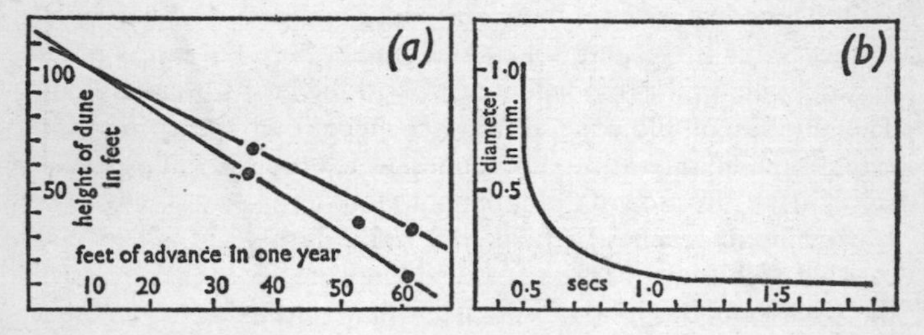


Fig. 95 (*a*) Dune-height and speed of advance (drawn from Beadnell's data)
(*b*) Time taken for spherical sand-grains to fall 4.3 inches through air (re-drawn after Bagnold)

lack of suitable wind-records, and there is little agreement about the precise connection between wind-direction and the alignment of seif dunes. Nevertheless it seems clear that seif dunes differ fundamentally from the sand-ridges which produce barchans, just as those sand-ridges differ from ripples. The differences are not merely differences of scale.

One possible origin of seifs is the operation of dune-building winds blowing from more than one direction. Fig. 94*b* illustrates the combination of barchan-building winds from the west with additional sand-driving winds from the southwest. Progressive elongation of one horn of the barchan enables it to collect sand and to grow still further. Alternatively, seifs might be constructed along the resultant direction of two sets of oblique winds, roughly equal in duo-building power. But neither possibility seems readily to explain huge regional seif systems, such as the extensive seif spirals of the Australian deserts.

Because they grow so much higher than do barchans, seifs are the more effective in trapping coarse sand, which can accumulate thickly on their lower flanks – hence the view that whalebacks of coarse material are the relics of former seifs. Occasional barchans are found in the troughs between seif ridges, in all probability because the flow of air is firmly guided by the flanking walls. Thus the unidirectional winds necessary to build barchans can blow locally in regions where the seif is the characteristic dune-type.

On some exposed sandy shores, yet another type of dune is formed. This is the parabolic or U-dune, where the horns point *up*wind – i.e. in the opposite direction to the horns of a barchan. The process of blowout appears significant in the formation, extension, and migration of U-dunes, which typically align themselves in the direction of the dominant wind.

Moving air reaches high speeds, and in terms of bulk load is capable of shifting quantities of dust. The capacity of the air above the Mississippi Basin to transport solid material has been calculated as 1000 times as great as the capacity of the Mississippi river-system. Although the tranporting capacity of the air is never fully realized in practice, concentrations of dust in major storms are known to reach 200,000 tons per square mile of land-surface.

Deposition is equally impressive, outside the desert borders as well as within them. At diameters of 0·08 mm. and where the grains are approaching or entering the grade of silt, particles can stay suspended in turbulent air. They fall very slowly even when the air is still, like solid particles settling in water (Fig. 95*b*). The finest particles can remain aloft for very long periods. This is why dust rises to great heights and travels long distances, while sand proper is confined to low levels and makes but short flights. Where thick sheets of wind-laid dust accumulate, the deposit is styled *loess*.

The principal region where loess is now forming lies in the upper Hoangho Basin, within the range of thick deposition by winds off the Gobi Desert. But with the northwest monsoon of each winter season, yellow-brown dust is borne as far as Pekin. Once deposited, loess can resist erosion. Its average grain-size is 0·05 mm. – small enough to make the surface smooth in response to blowing winds, and small enough to enable the particles to bind together under the influence of absorbed water.

*

Desert margins and desert borderlands commonly show signs of different climates in the past. Former increased dryness is well documented by dead dune systems outside existing deserts: linear dunes from the Kalahari once reached to the lower Congo. In addition, many dune systems inside existing deserts are also dead. Former increased wetness is proven by high beaches, formed by permanent bodies of water in basins which are now dry. We are forced to conclude that desert margins have fluctuated, both inside and outside existing limits. Documentation is so far most abundant for the last 15,000 years or so. At the beginning of this period, deserts were being pressed back by increased rain. Later, however, the climates of desert borders have been even drier than they are today. Discounting the undeniable fact that man can make deserts spread – as for instance through the uncontrolled grazing of the destructive goat – we infer that the climatic balance of existing margins is an uneasy one. The prospect seems to be a desert advance or a desert retreat. Only the future can tell which.

CHAPTER SEVENTEEN

# *Instances and Implications*

ANY problem of earth-sculpture which is at all complex involves problems belonging to other sciences. Differences among deserts, icefields, and river-basins are essentially climatic. Uplift which begins an erosion-cycle is a matter for geologists. Thus even when the sculptural history of a particular piece of landscape has been satisfactorily reconstructed, important questions may remain outstanding. In this chapter, examples will be given of the interlocking of problems and of the implications of certain studies of erosional history.

A hundred and sixty miles east of Paris stands the town of Toul. Overlooking the river Moselle, and commanding a deep gap through a line of hills, Toul occupies a highly strategic position. It has been repeatedly attacked, defended, taken, and retaken. The Toul Gap is as interesting scientifically as it is strategically, for it marks the former course of the Moselle towards the Meuse.

This is a classic site in the study of earth-sculpture. The evidence for the old route of the Moselle through the gap is very clear indeed. Not only does the river make directly towards the gap, before turning abruptly away to the northeast, but gravel deposits run down the valley of the river and into the opening. Furthermore, the gap is carved in bold windings – the incised meanders of the river which no longer passes through it.

At first sight, the interpretation of the Toul Gap seems simple in the extreme. The Moselle above Toul appears to have been captured by a tributary of the Meurthe. The Meurthe, flowing over weak rocks and cutting rapidly downwards, developed tributaries which likewise deepened their valleys with some speed. One of these tributaries, working back into the valley of the upper Moselle, appears to have captured its waters, leading them towards the Meurthe (Fig. 96). On the western side of the hills, beyond the gap, the Meuse shows obvious signs of shrinkage. No less would be expected, for it has been deprived of more than half its former catchment. It seems self-evident that the shrinkage is the result of river-capture.

Such a reconstruction can be made, on sight, from a map. It has all the merit of simplicity, and is all the more attractive because

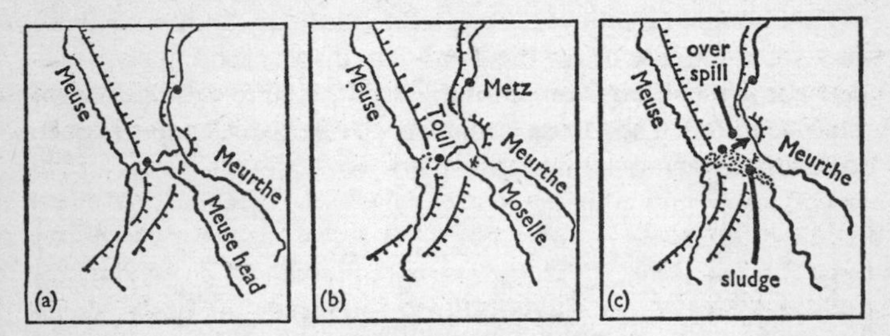


Fig. 96 Diversion of the Moselle: a→b, according to Davis; a→c→b, according to Tricart

river-capture is likely to occur in terrain of this kind. But the reconstruction, for all its simplicity, has the defects of serious error. The true facts of the case are more complicated, but at the same time highly instructive. The modern interpretation of the Toul Gap illuminates a general and important problem.

It is now known that the diversion of the upper Meuse from its path through the gap cannot be classed as river-capture at all – at least in the usual sense of the term. The diversion was due not to the piratical success of a tributary of the Meurthe, but to the effects of very cold climate. During one of the four glacial maxima of the Ice Age, the valley of the upper Moselle received great quantities of rock-waste, which sludged down the valley-sides over the frozen ground beneath (p. 193). Although part of the sludge was carried away by the river, there was too much of it for the valley to be kept clear, and some infilling took place, particularly in the wide valley upstream of the narrow Toul Gap, where the river wandered brokenly over a wide spread of sand and gravel. So deep did the fill become that the Moselle, raised to higher and higher levels, eventually spilled sideways across a low pass into the basin of the Meurthe. Cutting sharply into its new bed, it fixed itself in the course which it has followed ever since. (Fig. 96, a→c→b.)

By this diversion, the Meuse below the western end of the Toul

Gap lost rather more than half its catchment area and rather more than half its volume. Now since there is a connection between the volume of a river and the size of its channel, and between the size of the channel and the dimensions of meanders, a reduction in volume should be followed by a reduction in meander-size. Sure enough, the Meuse below the Toul Gap displays shrunken meanders. Its present loops are contained in the large windings of its valley (Fig. 97). The Meuse is a *misfit* – a river which appears once to have been far larger than it is today.

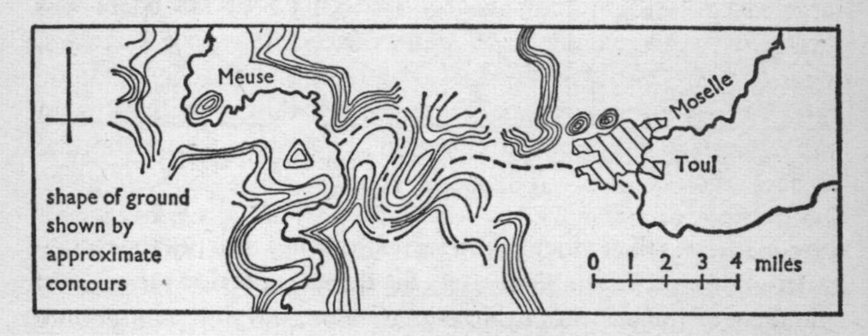


Fig. 97 The Toul Gap and the misfit Meuse

The re-interpretation of the diversion of the upper Moselle from the Meuse does not affect the conclusion that the Meuse was reduced in volume by the loss of headstreams. But as the Meuse lost, so the Moselle–Meurthe system gained. It should be possible to obtain evidence of an increase in volume on the one side, corresponding to the decrease on the other.

But when the various rivers are inspected, it is at once seen that the Moselle is, like the Meuse, a misfit. The Meuse *above* the Toul Gap is also a misfit (Fig. 97). In both cases, the meanders of the river are contained in the large curves of a winding valley. The Meuse above the gap cannot possibly have been reduced in volume by the diversion of the upper Moselle, while the lower Moselle should show signs not of reduced, but of increased volume. Tributary streams also possess winding valleys in addition to winding channels. Clearly, something more is involved than the mere diversion of the upper Moselle from its former course through the Toul Gap.

Before the argument is carried farther, the significance of winding valleys needs to be clarified. If the valley-windings are true meanders, there can be no doubt that they were cut by large ancestral streams. They were originally interpreted as true meanders solely in respect of their surface form. They have steep slopes on the outsides of bends and gentle slopes on the insides – slopes similar in all respects to those cut by an incised meandering river.

The view that valley-windings are true meanders can, in fact, be readily confirmed. The windings do indicate the former presence of large rivers. But so long as information about meanders was limited to the general observation that big rivers have big meanders, while little rivers have little ones, the full implications of winding valleys could not be realized. These valleys were freely interpreted as the result of river-capture. Dismembered streams were thought to have been suddenly reduced in volume, and to have adjusted their traces accordingly. It was largely by accident that misfit streams were first recognized in terrain where capture was likely to have occurred. The hypothesis of reduction by capture gained widespread favour, the Meuse being often quoted as a stream reduced in volume by the loss of its headwaters.

However, when the investigation of misfit streams in general, and of the Meuse in particular, is given some kind of factual basis, doubts about the influence of diversion begin to appear. For instance, it can be shown that on many misfit rivers the wavelength of valley-windings is about 10 times the wavelength of existing meanders. A reduction in wavelength to $\frac{1}{10}$ of the original figure implies a reduction in volume by about 99 per cent. If volume is to be reduced in this proportion by capture, it must be supposed that 99 per cent of the original drainage-basin is lost.

This is not the only objection to the idea of reduction by capture. In many parts of Europe, the U.S.A., and western Russia, as well as in New Zealand, all or most of the sizeable rivers are misfits. It is quite impossible to imagine any system of captures capable of reducing the volume of every stream throughout a whole region. Moreover, the degree of reduction in volume, as indicated by the ratio of wavelength between valley-windings and existing meanders, tends to be constant in any given area. It must be concluded that what has affected one river has affected all.

Thus the misfit condition of the Moselle, of the Meuse above the Toul Gap, and of their tributaries, becomes part of a general problem. Although the Meuse below the gap has undeniably been reduced by the diversion of the upper Moselle, its shrinkage through this cause is incidental to a general shrinkage of all the streams in the district. The special case of the diversion of the upper Moselle provides no help in solving the problem set by misfit streams as a group. The capture hypothesis is to be abandoned as a general solution, and additional information sought in the field.

Now if the valley-windings are true meanders, as from their shape they appear to be, it seems possible that some trace might remain of the beds of the large streams which once eroded them. When bore-holes are sunk into the flood-plains, it is found that the alluvium is contained in large meandering channels – channels which wind round the valley-bends. Here are the beds of the ancestral streams which cut the winding valleys. Significantly, they are about 10 times as wide as the present river-beds, just as the valley meanders are about 10 times as long as the meanders of the existing streams. The origin of misfit streams is thus to be interpreted in the manner shown in Fig. 98.

**Fig. 98 Development of misfit streams**

Part of the relevant field-work has been done in the Cotswolds, which are noted for their impressive array of winding valleys (Plate 29). Some of the Cotswold rivers are known to have lost their headwaters by river-capture, while others have certainly not. It is found that all the rivers, whether captured or not, are misfit, and that they are misfit to the same degree. The general shrinkage which reduced them to their misfit condition was therefore independent of capture, and was later than any captures which may have occurred. This conclusion is confirmed by the fact that the possible capturing rivers are also misfits, and that they are just as much misfit as the deprived Cotswold streams. These observations point directly to a general change in runoff as the cause of a misfit condition.

The possible suggestion that valley-windings are cut when the river is in flood can be immediately dismissed. Observation shows that high floods, such as those to be expected once in 100 years, fail to shift the alluvium which fills the old stream-beds. In any case, stream-channels are shaped not by flood-water but by the discharges which occur about once a year. What is needed is more water at all times. The lost water cannot merely have sunk into the ground, for some misfit streams are underlain by impermeable bedrock. It cannot have been provided by glaciers or ice-dammed lakes, for misfit streams occur well beyond the extreme limits reached by the former ice-sheets. The only possibility is a change of climate.

For some quite inexplicable reason, the great changes of temperature which accompanied glaciation are accepted without question, but suggestions about changes in rainfall are looked on with suspicion. It can, of course, be urged that the large volumes of water needed to erode the great winding channels were provided by melting snow. This might have been so, if only the channels had been cut at times of cold climate. But even in that event, the erosion of the channels would depend on a climate different from that of today. As it is, however, some of the channels are known to have been cut when the climate was warmer, not colder, than it now is. We must, therefore, appeal to high rainfall at some former time.

Indications are that high channel-forming discharges were experienced at intervals during the Ice Age, and that the last general occurrence was about 10,000 years ago during waning glaciation. Low temperatures at that time do not wholly explain the great runoff: increased precipitation is also needed, up to one-and-a-half to twice as much as the precipitation of today. The conditions envisaged could increase channel-forming discharge to between 20 and 60 times its present values – enough to scour the large channels and to excavate valley meanders. The changes in precipitation appear to correspond to changes in the tracks, frequency, and strength of travelling lows. At this point the study of earth-sculpture merges into the mathematical and astronomical study of past climates – a topic too vast to be explored here. Enough has probably been said, however, to show that the landscapes observed today have not necessarily been produced in present-day conditions.

*

The piece of landscape-history next to be reconstructed well illustrates the great use which can be made of a single clue. It concerns the origins, development, and disruption of the ancestral Trent. In its present form, the Trent rises in the southwestern Pennines, flows roughly eastwards to Nottingham, and then turns northwards before emptying into the Humber. Its original course was very different from this.

The vital clue lies beyond the limits of the basin of the existing Trent. It consists of the well-known winding gorge, traversed by the Dee and containing the small town of Llangollen (Fig. 99).

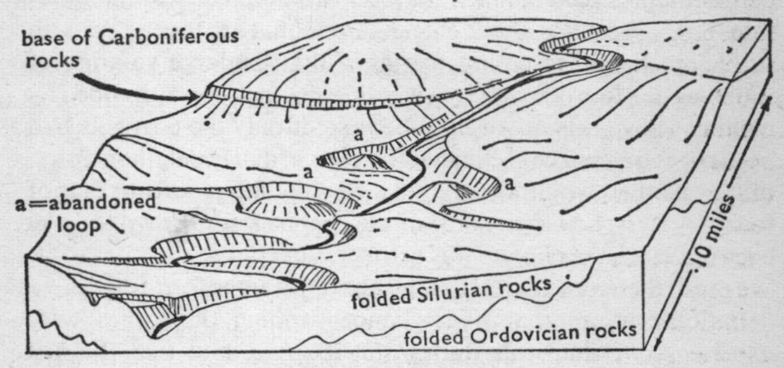


Fig. 99 The Dee gorge

Seen from the higher slopes or from the air, the gorge is quite impressive as a landscape-feature. When it is considered in relation to the pattern of the surrounding country, it becomes even more remarkable. It cuts through a long line of substantial hills which, unbroken except at this point, run diagonally across the whole width of Wales, including Cader Idris at one end and the Berwyn Mountains near the other.

Aligned from southwest to northeast, the hills reflect the structural grain which was impressed on most of Wales by an ancient episode of mountain-building. The gorge, transecting the structural lines at a high angle, belongs to a river-course which is independent of structure. It can be concluded that the river has dug down from some vanished cover into the rocks which are now exposed, or at least that it once flowed over a very gentle slope cut indiscriminately across belts of strong rock and lines of weakness. In the early landscape the underlying structural pattern was not expressed, for the existing valleys had still to be eroded.

At the side of the gorge, the ground rises to heights of 1800 feet. On the floor, heights of some 300 feet are recorded. A river has been following this line throughout the whole of the time taken to cut a trench 1500 feet in depth. It remains to discover where that river rose, and where it emptied into the sea.

The due west-to-east line of the gorge suggests that the old head-waters might have risen in the high ground of North Wales, which reaches heights of 3000 feet or so. Indeed, there is no need to look any farther for the ancient gathering-grounds. A single large valley-head can be reconstructed in the Snowdon district, when allowance is made for glacial breaches in the existing watershed (Chap. 14). The restored watershed curves along the line of crests, opening only on the eastern side (Fig. 100). The reconstructed head-valley – the valley-head of the ancestral Trent – is aligned directly on the Dee gorge.

East of the gorge, the land is based on weak and easily eroded rocks. Piratical streams draining to the Irish Sea, and working back into these rocks, had succeeded in dismembering the old trunk river before the onset of glaciation. Chief among them was the Dee, which tapped the main stream to the east of the gorge, diverting its waters to the northwest. Additional piracies west of the gorge were carried out by the Conway and the Clwyd. The Clwyd had the advantage of working headwards along a down-sunken trough filled with weak sediments, while the Conway extended itself along weak structures in rocks which are generally strong. The Clwyd displaced the watershed by capturing tributaries, while the Conway worked right back across the line of the main stream.

Meanwhile, the latter had thrown out tributaries along the structural grain. One of these is now represented by the upper Severn, a second by the Vyrnwy, and a third by the Dee below Lake

Bala. These streams, being so well accommodated to the structure of the country, are not considered original elements of the drainage net. Instead, they are regarded as having developed outwards from the trunk river, following lines of weakness.

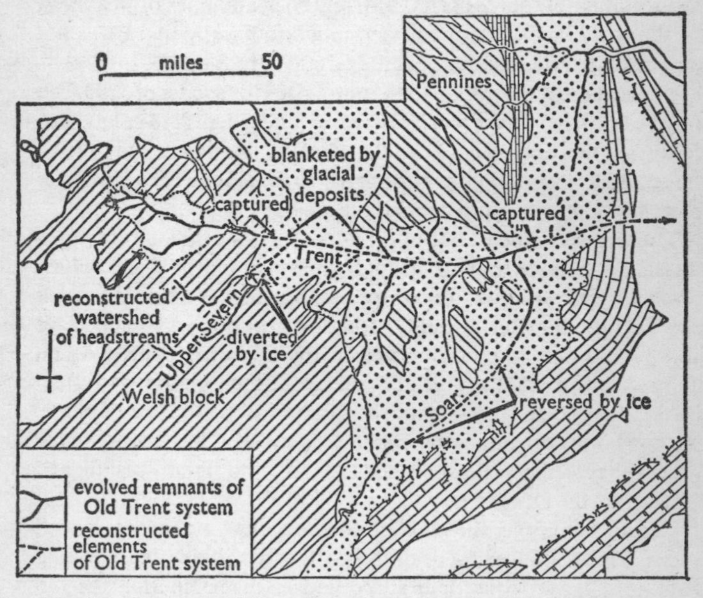


Fig. 100 The ancestral Trent (based on the work of Linton, Shotton, Dury, and others)

In addition to being itself dismembered, the ancestral Trent was to lose the Vyrnwy and the upper Severn. These were diverted during the Ice Age. During one of the glacial maxima – the last of the four – a lobe of ice invaded the lower Dee valley from the direction of the Irish Sea. It blanketed the lower course of the Dee with sand, gravel, and boulder clay, without preventing that river from resuming a course towards the Irish Sea when deglaciation occurred. On its southern side the ice pressed against lines of low hills, damming up the waters of the Vyrnwy and upper Severn in the form of lakes, and forcing them to spill across passes to the south (Fig. 101). One pass, on the site of Ironbridge, was so deeply eroded that it still provides a route for the waters of the Severn. The upper Severn and its tributary the Vyrnwy have been permanently diverted from the old Trent system.

In these various ways, the ancestral Trent was deprived of every one of its headstreams. It might, therefore, seem difficult to recon-

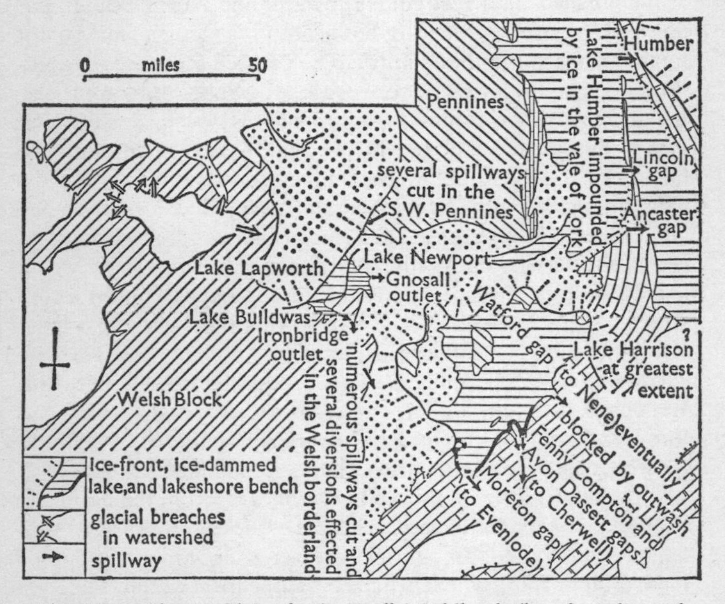


Fig. 101 Ice-dammed lakes in the English Midlands (based on the work of Shotton, Dury, Wills, Raistrick, and others)

struct the former river, when so few of its branches remain. Nevertheless, there is very good evidence for its existence in the middle section, as will now be shown.

It has been explained in Chap. 14 that the present Warwickshire Avon is a reversed stream, which in pre-glacial times used to flow towards the northeast, instead of towards the southwest as it does today. The older and longer Soar, which the Avon has replaced, was once tributary to the original Trent. Other former tributaries can be identified, flowing towards the south, in the southern

Pennines. Like the Dee in the Dee gorge, these rivers have cut deeply down across belts of resistant rock. The reconstructed Trent provided an outlet for the waters of the south Pennine streams as well as for those of the old Soar (Fig. 100).

Farther east still, beyond the line of Nottingham, the early Trent pursued its direct course towards the North Sea. It was diverted by a river which worked headwards along a wide belt of weak rocks around the Humber. The detailed sequence of events was complex, involving not only successive river-capture but also the existence of ice-dammed lakes in what is now the lower Trent valley. For the sake of brevity we may pass over the intricate history of the Trent below Nottingham, merely noting that the gaps through Lincoln Edge at Ancaster and at Lincoln once carried large rivers or large quantities of glacial meltwater, becoming converted into deep notches in the solid line of hills.

By a process of piecemeal restoration, a river has been reconstructed rising in North Wales and flowing right across England. Reconstructions of other early streams, such as the Kennet–Thames (p. 184), the Tyne, the Tweed, and streams in the Scottish Highlands, indicate parallel courses from west to east. It can therefore be inferred that the original rivers were initiated by uplift in the west, depression in the east, or by both combined.

Although so little of the west–east lines is left in the drainage-network of today, the general principle of the restorations can confidently be accepted. Two further problems then arise – the date of the crustal warping which provided a slope from west to east, and the nature of the land on which the eastward-flowing rivers first developed. The following discussion of these problems is not meant to give a complete and final solution, but merely to illustrate one of the possible modes of reasoning.

The last sea which spread widely across the land of Britain was that in which the Chalk was laid down. About 100 million years ago began a long period of subsidence, when the land was invaded by the sea and the distinctive sediments of the Chalk were laid down upon the sea-bed. Although it cannot be said with certainty how far inland the Chalk sea came, it is at least certain that any later marine invasions were confined to the Hampshire Basin, the London Basin, and the eastern part of East Anglia. The slowly extending Chalk sea was the latest which could possibly have submerged the high ground of the north and west.

It is obvious that the Chalk was formerly more extensive than it is today. The Lincolnshire Wolds, the Chilterns, and the Berkshire Downs have been eroded back to their present lines. Looking forward in time one can imagine them in full retreat, being cut back farther and farther still towards the east and south. But if the apparent flow of time is reversed, so that one looks backward into landscape-history, the Chalk hills can be imagined as advancing towards the west and north. The question is, how far westwards and northwards did they once extend?

Chalk is preserved beneath the lava-sheets of Antrim, where it can be inspected on the flanks of the glens. Chalk is probably present beneath the Irish Sea, off the coast of Lancashire, for flints derived from Chalk have been carried by upward-moving ice on to the eastern side of Anglesey. It is tempting to project the Chalk of Lincolnshire over the Pennines, joining it to the probable Chalk in the Irish Sea Basin and to the known Chalk of Antrim. It could then be supposed that the early rivers of Great Britain – the ancestral Trent among them – were initiated on a broad sheet of Chalk which, spread generally over the country, became land as a result of upwarping towards the west.

The restoration just outlined represents a view which was in fashion about 50 years ago, and has again attracted notice recently. However, it is generally conceded that nothing can remain of the eroded surface on which a general cover of Chalk may once have lain. The highest summits of Wales and Scotland stand too low to form part of the uplifted and now-destroyed floor of the Chalk sea.

A more modest – and more probable – view is that the Chalk thinned away against the high ground, its surface merging imperceptibly into the planed-off surfaces of the older rocks. In either case, uplift in the west has still to be described and accounted for. It is required in explaining the origin of the master streams which once flowed from west to east across the country.

The uplift must be fitted into geological history, coming later than the deposition of the Chalk. That is to say, it must be placed in the Tertiary division of geological time.

There is abundant evidence of crustal disturbance, during the

Tertiary period, in the northern and western districts. Great fissures opened in the north of Ireland, emitting highly mobile lava which solidified as sheets of basalt. Volcanoes erupted off the west of Scotland, on Skye, Mull, and Arran, where they steamed, rumbled and exploded, poured out streams of lava, showered down dust, and built tall cones. In their full vigour, these volcanoes rose thousands of feet above their surroundings. Now, long extinct, they have been reduced to ruins by the inescapable force of erosion.

Contrasts between the manner of volcanic activity between Antrim and the Hebrides are less significant, in the present context, than the record of eruptions from both sides of the Irish Sea Basin. This basin has probably been formed by crustal subsidence. Taken in combination, the depression of a strip of crust, the volcanic outbursts in adjacent districts, and the easterly slope required to account for the early British rivers are highly suggestive. They make it possible to conclude that the ancestral Trent was initiated, the floor of the Chalk sea upraised, and the Irish Sea depressed by a distinctive type crustal movement – namely, the upwarping of an elliptical dome (Fig. 102).

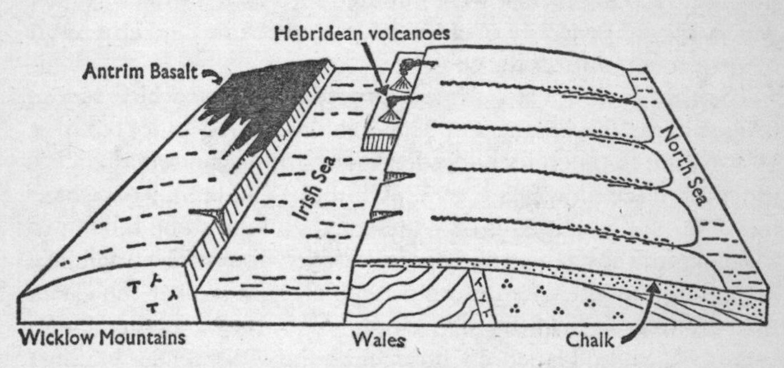


**Fig. 102 The British area as a Cloosian dome (diagrammatic)**

The eminent geologist Hans Cloos maintains that uplifts of this sort have occurred in several parts of the world. For instance, the huge valleys which containt he narrow lakes of East Africa represent, according to Cloos, depressed strips running across the centre of a dome. Significantly, volcanic outbreaks still occur on the margins of the East African troughs. Cloos claims the rift valley of the Rhine as a second example of the central depression, and the Irish Sea as a third.

Since Cloos was concerned with the theory of crustal deformation and not with the erosional history of landscape, a discussion of his work would be out of place here. But it should be said that his formal scheme accords very well with what has been discovered of the origins of British drainage. Field studies throughout the whole of Great Britain indicate that the original watershed lay close to the western edge of the present land, and that the original rivers flowed directly to the east down a long slope which can only have been produced by crustal movement.

*

This last case, beginning as historical geomorphology and thus being, by definition, timebound, leads far into a discussion of earth-movement by which the time-scale is, at the least, greatly extended. The earlier case of misfit streams, where field investigation was initially historical, is deeply involved in practice with hydrologic theory and with general hypotheses of climatic change. As was pointed out earlier, to accept climatic change as inevitable encourages open-system thinking: and the open-system approach is especially suited to intensive work on small areas, and to work on processes in operation.

Climatic change, continental drift, and open-system study combine – as it were by accident – in current work in Australia. A vast number of Australian residuals, including some of considerable extent, are capped by surface crusts. The generic name for the crust is *duricrust* – a name for which I have no responsibility whatever, although the coincidence seems happy. Along much of the eastern coastland of Australia, and in the bordering mountains, the duricrust takes the form of laterite. It is composed largely of compounds of iron and aluminium. In great expanses of the inland, the crust is silcrete – its main constituent is silica. Thicknesses of crust as great as 40 feet are on record.

Below the crust comes a mottled zone, grossly blotched by iron stains. Below the mottled zone lies, in turn, a pallid (bleached) zone, where in some sections the original rocks have been wholly converted to kaolin. The combination of crust, mottled zone, and pallid zone constitutes a deep-weathering profile, which can attain a total thickness of 250 feet. Here is the profile of a former soil, developed on an enormous scale.

A widely held view is that the pallid zone, formerly lying permanently below a water-table, was perennially subjected to chemical reduction. The mottled zone experienced seasonal rises and falls of the water-table, while above the highest seasonal level of underground saturation, compounds of iron, aluminium, or silicon were selectively concentrated. These formed the crust (= indurated zone), in conditions and as a result of processes which are not yet fully known. Climate inferred from deep weathering and duricrusting differs markedly from the climate of today, especially in dry inland parts where (artesian water apart) underground water does not exist. Areas which are now desert or semi-desert appear formerly to have experienced pronounced seasonal rain. Furthermore, comparison of the former and contemporary results of weathering in the humid coastlands suggests that deep weathering must have been promoted also by temperatures higher than those of today.

Little is yet known for certain about the date of deep weathering and of duricrusting; but the most likely age of both seems to be of the order of 25 million years – somewhere in the Oligocene and/or the Miocene. Oxygen-isotope analysis of fossil shells demonstrates that, at that time, temperatures in southeastern Australia at least were much higher than they are now. A date of this order implies that the Australian landmass, when it was being deeply weathered, occupied some other than its present position. It has been displaced by continental drift – as, too, has Africa, in parts of which deep weathering and duricrusting are also common. Reconstruction of the former climate, beyond what is possible from the scanty clues detected so far, is made all the more difficult because the general circulation of the atmosphere, at the relevant time, may have been very different from what it now is, at a time when glaciation is by no means over.

Open-system study comes into consideration, in relation to the destruction and removal of surface crusts. Crusted residuals supply parts of the stony spreads which form the distinctive gibbers (the so-called gravel deserts) of the Australian inland. Where a residual in dry country carries a silcrete capping, the particles shed from the typically scarped edge of the cap can be identified far downslope. They produce an abundant surface litter down the constant slopes, and far out upon the fringing pediments. Changes in their size and shape, between positions high upslope and positions low downslope, can readily be defined: and these changes must necessarily relate to vertical lowering and to lateral displacement. Information so far available suggests that the main changes are changes in size. The moving particles become progressively reduced in length, breadth, and thickness, but the proportion of one dimension to another does not greatly alter, except that edges may become progressively rounded.

In the present context, the chief point is that the ultimate origin of the surface crust, however interesting in itself, has no bearing on studies of change during transport. The investigated slopes are looked on as working (open) systems, with material entering at the top of a given length, passing through, and being lost at the bottom. Inquiry into the origin of surface crusts is designed to answer the question, *What happened?* Investigation of particle-shape, particle-size, and of the relation of size and shape to slope refers to the question, *What happens?* – i.e. *What is happening now?* There follows inevitably from this latter question, the further question, *Why?*

CHAPTER EIGHTEEN

# *Conclusion*

A BEGINNER in geomorphology needs to solve problems of scale, of perspective, of map-convention, and of vocabulary. Land-forms in the field rarely display themselves in simple panoramas, being seen in perspective and not in plan. Whereas landscape-pattern and its relationship to structure seem abundantly clear in small diagrams, mere size and distance may, at the outset, defeat the inexperienced observer on the actual ground. High viewpoints, a strong visual and imaginative sense, and a sense of direction, prove enormously helpful. But some kind of small-scale representation is needed if observed features are to be reduced to order, and if their inter-relationship is to be defined. Such representation usually means a map. Relief maps of all kinds are highly conventional; morphological maps appear even more so, since they employ special symbols and differ strongly in general appearance from more familiar maps. However, morphological maps often reduce themselves in practice to distributional records of steep and gentle slopes. Where good base-maps already exist, morphological mapping is fairly simple – it begins with the drawing of lines along the boundary between steeply sloping and gently sloping facets of the ground. As soon as a pattern begins to emerge, the work is succeeding.

The foregoing text and its accompanying illustrations are meant to facilitate recognition of what may be called stock features of the landscape, and to attach to those features their correct names. But the best way to learn about landscape is to be shown actual pieces of country, and to have them explained on the spot. The only reasonable way to learn structural geology – a most useful accompaniment of geomorphology – is to attend a course of instruction. To those readers interested in extending their knowledge of land-forms I strongly commend indoor classes, walking over the ground, and conducted excursions. In Britain, the instructors and leaders required for classes and excursions may be had from universities, from local educational authorities, and from natural history societies.

Meanwhile, the following list of references is appended as a guide to further reading. The items cited vary much in level of difficulty; and a reader encountering one that is clearly far too technical should lay it aside without regret. With few exceptions, the general texts range very broadly through the fields of earth-sculpture and earth science. Individual papers are chiefly limited to single topics. Although the half-life of usefulness of papers is only ten to twenty years – that is to say, a paper loses about half its initial impact in about fifteen years, and three-quarters in about thirty years, and so on – references to individual papers which are fundamental, necessary, or both, have been retained.

A pervasive problem in selection is that works written by non-specialists for the popular readership risk being unsound, out-of-date, or both together. Particularly is this true of the treatment of rivers. A most remarkable instance of the opposite sort is Holmes's massive and authoritative introduction to physical geology. This, written for specialists, is so compelling that non-specialists find themselves unable to put it down. Among equally deliberate and equally successful efforts to communicate, those of Leopold and Langbein deserve special mention. These two innovators belong to an all-too-small group of outstanding researchers, who wish not only to explore new ideas, but to disseminate their findings to a general readership.

The present author concurs. Land-form study is exciting. Anyone liable to excitement by it deserves to be brought in.

# REFERENCES

## CHAPTER ONE

General works, also relevant to subsequent chapters or to parts of them:

R. J. Chorley (ed.), *Water, Earth, and Man*. London, 1969.
R. J. Chorley and P. Haggett (eds.), *Models in Geography*. London, 1967.
W. M. Davis, *Geographical Essays*. New York, 1909.
G. H. Dury (ed.), *Essays in Geomorphology*. London, 1966.
G. H. Dury, *Perspectives on Geomorphic Processes*. AAG, Washington D.C., 1969.
R. W. Fairbridge, *The Encyclopaedia of Geomorphology*. New York, 1968.
A. Holmes, *Principles of Physical Geology*. Edinburgh and London, 1965.
L. D. Leet and S. Judson, *Physical Geology*. Englewood Cliffs, N.J., 1965.
C. R. Longwell, R. F. Flint, and J. E. Sanders, *Physical Geology*. New York, 1969.
W. Penck, *Morphological Analysis of Landforms*. London, 1953.
W. D. Thornbury, *Principles of Geomorphology*. New York and London, 1969.

## CHAPTER TWO

E. M. Bridges, *World Soils*. Cambridge, 1970.
M. A. Carson, *The Mechanics of Erosion*. London, 1971.
I. Douglas, 'The Efficiency of Humid Tropical Denudation Systems'. *Trans. Inst. Brit. Geog*. **46**. 1969, 1.
C. Ollier, *Weathering*. New York, 1969.
D. B. Prior, N. Stephens, and D. R. Archer, 'Composite Mudflows on the Antrim Coast, etc.' *Geog. Annaler* **50** (A). 1968, 65.
J. Tricart, *The Landforms of the Humid Tropics, Forests, and Savannas*. London, 1972.
E. Yatsu, *Rock Control in Geomorphology*. Tokyo, 1966.

## CHAPTERS THREE AND FOUR

W. W. Bishop, 'The Pleistocene Geology and Geomorphology of Three Gaps, etc.' *Phil. Trans. Roy. Soc.* (B). 241, 255.

R. J. Chorley, 'Climate and Morphometry'. *Journ. Geol.* **65.** 1957, 628.

T. H. Clark and C. W. Stearn, *Geological Evolution of North America.* New York, 1968.

G. H. Dury, *Map Interpretation.* London, 1971.

A. Dwerryhouse, 'The Underground Waters of Northwest Yorkshire'. *Proc. Yorks. Geol. Soc.* New Ser. **15.** 1903, 5, 248.

R. E. Horton, 'Erosional Development of Streams, etc.' *Bull. Geol. Soc. Amer.* **56.** 1945, 275.

Institut Géographique National (pub.), *Relief Form Atlas.* Paris, 1952.

M. A. Melton, *An Analysis of Relations among . . . Climate, Surface Properties, and Geomorphology.* New York, 1957.

A. N. Strahler, 'Hypotheses of Stream Development in the Folded Appalachians, etc.' *Bull. Geol. Soc. Amer.* **56.** 1945, 45.

A. N. Strahler, 'Hypsometric (area-altitude) Analysis of Erosional Topography'. *Bull. Geol. Soc. Amer.* **63.** 1952, 1117.

A. N. Strahler, 'Quantitative Slope Analysis'. *Bull. Geol. Soc. Amer.* **67.** 1958, 571.

M. M. Sweeting, chapter in Dury (ed.), 1966, listed for Chapter One.

J. Thrailkill, 'Chemical and Hydrologic Factors in the Excavation of Limestone Caves'. *Bull. Geol. Soc. Amer.* **79.** 1968, 19.

M. J. Woldenburg, 'Horton's Laws Justified, etc.' *Bull. Geol. Soc. Amer.* **77.** 1966, 431.

M. J. Woldenburg, 'Spatial Order in Fluvial Systems, etc.' *Bull. Geol. Soc. Amer.* **80.** 1969, 97.

## CHAPTER FIVE

C. A. Cotton, *Volcanoes as Landscape Forms.* New Zealand, 1952.

C. A. Cotton, 'Geomorphic Evidence and . . . Transcurrent Faults in New Zealand'. *Révue de Géog. Physique et de Géol. Dynamique* (2) **1.** 1957, 16.

D. R. Crandell and H. W. Waldron, 'Volcanic Hazards in the Cascade Range', in *Focus on Environmental Geology* (ed. R. Tank). New York, 1973.

F. J. Fitch, 'The Development of the Beerenberg Volcano, Jan Mayen'. *Proc. Geol. Assoc.* **75.** 1964, 133.

A. Gunnarsson, *Volcanoes.* Reykjavik, 1973.

D. S. Hallacy, *Earthquakes*. Indianapolis, 1974.

W. Q. Kennedy, 'The Great Glen Fault'. *Quart. Journ. Geol. Soc.* **102.** 1946, 41.

K. C. McTaggart, 'The Mobility of Nuées Ardentes'. *Amer. Journ. Sci.* **258.** 1960, 369.

F. Press and D. Jackson, 'Alaskan Earthquake ... 1964 ...' *Science* **147.** 1965, 867.

## CHAPTER SIX

M. A. Carson and M. J. Kirkby, *Hillslope Form and Process*. Cambridge, 1972.

M. A. Carson and D. J. Petley, 'The Existence of Threshold Hillslopes, etc.' Trans. *Inst. Brit. Geog.* **49.** 1970, 71.

C. S. Carter and R. J. Chorley, 'Early Slope Development in an Expanding Stream System'. *Geol. Mag.* **98.** 1961, 117.

R. J. Chorley, *Geomorphology and General Systems Theory.* Washington D.C., 1962.

R. J. Chorley, 'The Application of Statistical Methods to Geomorphology'. In *Essays in Geomorphology* (ed. G. H. Dury), listed for Chapter One.

J. B. Dalrymple and others, 'An hypothetical nine unit landsurface model'. *Zeit. für Geomorph.* **12.** 1968, 6.

G. H. Dury and T. Langford-Smith, 'The Use of the Term Peneplain, etc.' *Austr. Journ. Sci.* **27.** 1964, 171.

J. T. Hack, 'Interpretation of Erosional Topography, etc.' *Amer. Journ. Sci.* **258-A.** 1960, 80.

C. A. M. King, 'Feedback Relationships in Geomorphology'. *Geog. Annaler* **52** (A). 1970, 147.

L. C. King, 'Canons of Landscape Evolution'. *Bull. Geol. Soc. Amer.* **64.** 1953, 721.

R. B. McConnell, 'Planation Surfaces in Guyana'. *Geog. Journ.* **134.** 1968, 506.

R. A. G. Savigear, 'The Analysis and Classification of Profile Forms'. Congrès et Colloque, Université de Liège, **40.** 1967, 271.

S. A. Schumm and R. W. Lichty', Time, Space, and Causality in Geomorphology'. *Amer. Journ. Sci.* **263.** 1965, 110.

M. Simons, 'The Morphological Analysis of Landforms ... the work of Walther Penck'. *Trans. Inst. Brit. Geog.* **31.** 1962, 1.

A. N. Strahler, 'Dimensional Analysis Applied to Fluvially Eroded Landforms'. *Bull Geol. Soc. Amer.* **69.** 1958, 279.

### CHAPTERS SEVEN AND EIGHT

G. H. Dury, 'Rivers and River Systems'. *Encyclopaedia Britannica* **15.** 1974, 874.

G. H. Dury (ed.), *Rivers and River Terraces*. London, 1970.

K. J. Gregory and D. E. Walling, *Drainage Basin Form and Process.* London, 1973.

W. B. Langbein and L. B. Leopold, 'Quasi-equilibrium States in Channel Morphology'. *Amer. Journ. Sci.* **262.** 1964, 782.

L. B. Leopold and W. B. Langbein, 'River Meanders'. *Scientific* American **214.** 1966, 60.

L. B. Leopold and T. Maddock, Jr, 'The Hydraulic Geometry of Stream Channels, etc.' *U.S. Geol. Survey Professional Paper* **252.** 1953.

L. B. Leopold and M. G. Wolman, 'River Channel Patterns'. *U.S. Geol. Survey Professional Paper* **292-B.** 1957.

L. B. Leopold, M. G. Wolman, and J. P. Miller, *Fluvial Processes in Geomorphology.* San Francisco and London, 1964.

A. A. Miller, 'Attainable Standards of Accuracy in the Determination of Pre-glacial Sea-levels'. *Journ. Geomorph.* **2.** 1939, 95.

S. S. Philbrick, 'What Future for Niagara Falls?' *Bull. Geol. Soc. Amer.* **85.** 1974, 91.

M. G. Wolman, 'The Natural Channel of Brandywine Creek, Pennsylvania'. *U.S. Geol. Survey Professional Paper* **271.** 1955.

M. G. Wolman and L. B. Leopold, 'River Flood Plains'. *U.S. Geol. Survey Professional Paper* **282-C.** 1957.

M. G. Wolman and J. P. Miller, 'Magnitude and Frequency of Forces in Geomorphic Processes'. *Journ. Geol.* **68.** 1960, 54.

### CHAPTERS NINE AND TEN

W. G. V. Balchin, 'The Erosion Surfaces of Exmoor, etc.' *Geog. Journ.* **117.** 1952, 453.

E. H. Brown, 'The Physique of Wales'. *Geog. Journ.* **123.** 1957, 208.

C. A. Cotton, 'Tests of a German Non-cyclic Theory and Classification of Coasts'. *Geog. Journ.* **120.** 1954, 353.

C. A. Cotton, 'Levels of Planation of Marine Benches'. *Zeit. für Geomorph.* **NF7.** 1963, 97.

J. L. Davies, 'Wave Refraction and the Evolution of Shoreline Curves'. *Geog. Studies* **5.** 1958, 1.

K. O. Emery, H. Niino, and B. Sullivan, 'Post-Pleistocene Levels of the East China Sea'. In *The Late Cenozoic Glacial Ages* (ed. K. K. Turekian). New Haven, 1971.

R. W. Fairbridge, 'Eustatic Changes in Sea Level'. *Phys. and Chem. of the Earth* **4.** 1961, 99.
A. Guilcher, *Coastal and Submarine Morphology*. London, 1958.
C. A. M. King, *Beaches and Coasts*. London, 1959.
V. J. May, 'The Retreat of Chalk Cliffs'. *Geog. Journ.* **137.** 1971, 203.
K. J. Mesolella and others, 'The Astronomical Theory of Climatic Change'. *Journ. Geol.* **77.** 1969, 250.
M. L. Schwartz, 'Seamounts as Sea-level Indicators'. *Bull. Geol. Soc. Amer.* **83.** 1972, 2975.
N. Stephens and F. M. Synge, 'Pleistocene Shorelines'. In *Essays in Geomorphology* (ed. G. H. Dury) listed for Chapter One.
M. J. Tooley, 'Sea-level Changes during the Last 9000 Years in North-west England'. *Geog. Journ.* **140.** 1974, 18.
R. A. R. Tricker, *Bores, Breakers, Waves, and Wakes*. London, 1964.

CHAPTER ELEVEN

V. V. Belousov, 'Modern Concepts of the Structure and Development of the Earth's Crust and the Upper Mantle of Continents'. *Quart. Journ. Geol. Soc.* **122.** 1966, 293.
E. C. Bullard, 'Continental Drift'. *Q. J. Geol. Soc. Lond.* **120.** 1964, 1.
S. W. Carey, 'The Rheid Concept in Geotectonics'. *J. Geol. Soc. Australia* **1.** 1953, 68.
C. Craddock, 'Antarctic Geology and Gondwanaland'. *Antarctic Journ. U.S.* **5.** 1970, 53.
W. Glen, *Continental Drift and Plate Tectonics*. Columbus (Ohio), 1975.
B. Gutenberg, 'The Energy of Earthquakes'. *Quart. Journ. Geol. Soc.* **112.** 1956, 1.
A. Hallam, *A Revolution in the Earth Sciences*. Oxford, 1973.
L. C. King, 'Pediplanation and Isotasy, etc.' *Quart. Journ. Geol. Soc.* **111.** 1955, 353.
R. B. McConnell, 'Geological Development of the Rift System of Eastern Africa'. *Bull. Geol. Soc. Amer.* **83.** 1972, 2549.
E. Orowan, 'Continental Drift and the Origin of Mountains'. *Science* **146.** 1964, 1003.
E. R. Oxburgh, *A Plain Man's Guide to Plate Tectonics*. London, 1975.
R. Stoneley, 'The Interior of the Earth'. *Adv. Sci.* **18.** 1961, 339.
J. Sutton, 'Development of the Continental Framework of the Atlantic'. *Proc. Geol. Assoc.* **79.** 1968, 275.
A. Wegener, *The Origin of Continents and Oceans*. London, 1924.
J. T. Wilson, 'Continental Drift'. *Scientific American* **208.** 1963, 86.

## CHAPTERS TWELVE AND THIRTEEN

C. Embleton and C. A. M. King, *Glacial and Periglacial Geomorphology*. London, 1968.

D. B. Ericson and others, 'The Pleistocene Epoch in Deep-Sea Sediments'. *Science* **146**. 1964, 723.

R. F. Flint, *Glacial and Pleistocene Geology*. New York, 1971.

W. L. Graf, 'The Geomorphology of the Glacial Valley Cross Section'. *Arctic and Alpine Research* **2**. 1970, 303.

M. F. Meier, *Mode of Flow of Saskatchewan Glacier, etc.* Washington D.C., 1960.

L. A. Neilson, 'Some Hypotheses on Surging Glaciers'. *Hull. Geol. Soc. Amer.* **79**. 1968, 1195.

E. J. Öpik, 'On the causes of ... Ice Ages ...' *Journ. Glaciology* **2**. 1953, 213.

C. B. Schultz and J. C. Frye (eds.), *Loess and Related Deposits of the World*. INQUA VII Congress, 1965.

F. W. Shotton, 'The Physical Background of Britain in the Pleistocene'. *Adv. Sci.* **19**. 1962, 1.

F. W. Shotton, 'The ... Methods of Absolute Dating within the Pleistocene ...' *Quart. Journ. Geol. Soc.* **122**. 1966, 357.

J. Steiner and E. Grillmair, 'Possible galactic causes for periodic and episodic glaciations'. *Bull. Geol. Soc. Amer.* **84**. 1973, 1003.

W. F. Tanner, 'Cause and Development of an Ice Age'. *Journ. Geol.* **73**. 1965, 413.

K. K. Turekian (ed.), *The Late Cenozoic Glacial Ages*. New Haven, 1971.

R. G. West, *Pleistocene Geology and Biology*. New York, 1968.

R. G. West and J. J. Donner, 'The Glaciations of East Anglia and ... Stone-Orientation Measurements, etc.' *Quart. Journ. Geol. Soc.* **112**. 1956, 69.

H. E. Wright (ed.), *Quaternary Geology and Climate*. Washington D.C., 1969.

## CHAPTER FOURTEEN

Book by Embleton and King, listed for Chapters Twelve and Thirteen·

E. Derbyshire, 'Fluvioglacial Erosion near Knob Lake, etc.' *Bull. Geol. Soc. Amer.* **73**. 1962, 1111.

G. H. Dury, 'Aspects of the Geomorphology of Slieve League Peninsula, Donegal'. *Proc. Geol. Assoc.* **75**. 1964, 445.

G. H. Dury, 'A 400-foot Bench in South-eastern Warwickshire'. *Proc. Geol. Assoc.* **62.** 1951, 167.

G. H. Dury, 'A Glacial Breach in the Northwestern Highlands'. *Scot. Geog. Mag.* **69.** 1953, 106.

G. H. Dury, 'Diversion of Drainage by Ice'. *Science News* **38.** 1955, 48.

D. L. Linton, 'Watershed Breaching by Ice in Scotland'. *Inst. Brit. Geog., Trans. & Papers*, 1949, 1.

D. L. Linton, 'Some Scottish River Captures Re-examined'. *Scot. Geog. Mag.* **65.** 1949, 123.

D. L. Linton, 'Some Scottish River Captures Re-examined – II'. *Scot. Geog. Mag.* **67.** 1951, 31.

D. L. Linton, 'The forms of Glacial Erosion'. *Trans. Inst. Brit. Geog.* **33.** 1963, 1.

H. E. Malde, *The Catastrophic ... Bonneville Flood, etc.* Washington D.C., 1969.

M. Poznansky, 'The Glacial Succession in the Middle Trent Basin'. *Proc. Geol. Assoc.* **71.** 1960, 285.

F. W. Shotton, 'The Pleistocene Deposits of the Area between Coventry, Rugby, etc.' *Phil. Trans. Roy. Soc.* Ser. B, **237.** 1953, 209.

## CHAPTER FIFTEEN

Books by Flint, West, and Embleton and King, listed for Chapters Twelve and Thirteen.

R. F. Black and W. L. Barksdale, 'Oriented Lakes of Northern Alaska'. *Journ. Geol.* **57.** 1949, 105.

R. J. E. Brown, 'Permafrost in the Canadian Arctic Archipelago'. *Zeit. für Geomorph.* Suppl. Bd. **13.** 1972, 103.

K. Bryan, 'Cryopedology – The Study of Frozen Ground, etc.' *Amer. Journ. Sci.* **244.** 1946, 622.

G. A. Kellaway and J. H. Taylor, 'Early Stages in the ... Evolution of ... the East Midlands'. *Quart. Journ. Geol. Soc.* **108.** 1952, 343.

J. R. Mackay, 'The World of Underground Ice'. *Annals, Assoc. Amer. Geog.* **62.** 1972, 1.

A. Rapp, 'Recent Development of Mountain Slopes in Kärkevagge, etc.' *Geog. Annaler*, **42.** 1960, 65.

P. J. Williams, 'Climatic Factors Controlling ... Frozen Ground Phenomena'. *Geog. Annaler*, **43.** 1961, 339.

### CHAPTER SIXTEEN

R. A. Bagnold, *The Physics of Blown Sand and Desert Dunes.* London, 1941.

G. H. Dury, 'Morphometry of Gibber Gravel, etc.' *Journ. Geol. Soc., Australia* **16.** 1970, 655.

G. H. Dury, 'Paleohydrologic Implications of some Pluvial Lakes, etc.' *Bull. Geol. Soc. Amer.* **84.** 1973, 3663.

A. S. Goudie, B. Allchin, and K. T. M. Hedge, 'The Former Extensions of the Great Indian Sand Desert'. *Geog. Journ.* **139.** 1973, 242.

L. K. Lustig, *Inventory of Research ... On Desert Environments.* Tucson, 1967.

G. E. Williams, 'The Central Australian Stream Floods of February–March 1967'. *Journ. Hydrology* **11.** 1970, 185.

### CHAPTER SEVENTEEN

H. Cloos, '*Hebung – Spaltung – Vulkanismus*'. *Geol. Rundschau* **30.** 1939, Zwischenheft 4A.

W. M. Davis, '*La Seine, la Meuse, et la Moselle*'. *Ann. de Géog.* **6.** 1896, 25.

G. H. Dury, '*La Meuse, rivière sous-adaptée*'. *Rev. de Géomorphologie Dynamique*, nos. **11–12.** 1956, 161.

G. H. Dury, *Principles of Underfit Streams.* Washington D.C., 1964.

G. H. Dury, *Subsurface Exploration and Chronology of Underfit Streams.* Washington D.C., 1964.

G. H. Dury, *Theoretical Implications of Underfit Streams.* Washington D.C., 1965.

G. H. Dury, 'Rational Descriptive Classification of Duricrusts'. *Earth Sci. Journ.* **3.** 1969, 77.

G. H. Dury, 'Relict Deep Weathering and Duricrusting in Relation to ... Paleoenvironments ...' *Geog. Journ.* **137.** 1971, 511.

G. H. Dury, 'Duricrusts'. *Encyclopaedia Britannica* **5.** 1974, 1008.

T. Langford-Smith, 'The Dead River Systems of the Murrumbidgee'. *Geog. Review* **50.** 1960, 368.

D. L. Linton, 'Midland Drainage'. *Adv. Sci.* **7.** 1951, 449.

A. E. M. Nairn (ed.), *Problems in Palaeoclimatology.* London, 1964.

# INDEX

*a Penguin Reference Book*

# A DICTIONARY OF GEOLOGY

D. G. A. WHITTEN with J. R. V. BROOKS

Geology – a science in which the amateur still counts for something – is an expanding world with a swelling list of fresh terms.

A new dictionary of geology is certainly needed and this one can now claim to be the most up-to-date compilation in English at a popular price.

With its invaluable table of minerals and useful bibliography, this dictionary is likely to appeal to students of geology, whether at school or college, and to scientists in other fields no less than to the prospecting army of amateurs in local societies or on 'further education'.

*a Penguin Reference Book*

# A DICTIONARY OF GEOGRAPHY

W. G. MOORE

This dictionary – now revised and enlarged – describes and explains such commonly met terms as the Trough of Low Pressure (from the weather forecast), a Mackerel Sky, a Tornado, the Spring Tides and hundreds of others. But there are also sections on such stranger phenomena as the Willy-willy of Australia, the Doctor of West Africa, the Plum Rains of Japan, the Volcanic Bomb, the Anti-Trades, the Bad Lands, and the Celestial Equator.

Because geography is largely a synthetic subject, the items of the dictionary are derived from many sciences, including geology, meteorology, climatology, astronomy, anthropology, biology. Even the most abstruse terms, however, are of the kind that the student is likely to meet in the course of his reading – terms that the author of geographical works employs but often has no space to define. The dictionary may thus help to clarify and systematize the reader's knowledge.